INTRODUCTION

It is now possible, provided you can afford the fare, to travel to various points on the earth's surface at a speed considerably faster than that of sound. At the turn of the century the only ways in which man could fly were by means of a glider or a balloon. Consequently – and no doubt rightly – the feats of the plane makers are rewarded with a kind of awed admiration. So much, it seems, has been accomplished in such a short time.

Such achievements have tended to turn popular imagination away from progress on that other element, the sea. The result is that a safe passage is taken for granted. When holiday-makers set off on a cruise, they do not consider the possibility of becoming shipwrecked. They seldom question whether such-and-such is a safe vessel, whether her captain is an able navigator, and so on. They give the matter no more thought than they would the prospect of a train journey.

Nevertheless, progress towards safety on the sea is more recent than one thinks. It was only at the beginning of the last century that a small vessel named the *Charlotte Dundas* demonstrated that steam engines were a practical method of propelling ships. The Royal National Lifeboat Institution, without which many lives would still be lost, was founded in 1823; and, until 1850, it was not necessary for the captains and other officers of a merchantman to be examined on their competence. Finally, it is only just over a hundred years since Samuel Plimsoll established a mark on the side of a vessel's hull to ensure that she was not overloaded.

Even in the middle of the nineteenth century, the score of shipwrecks around the coast of Britain was considerable. According to one contemporary writer (F. Whymper, author of *The Sea*), the average for the years 1852-71 worked out at the extraordinary figure of 1,446 per annum. Some of them, to be sure, were small craft, but by no means all of them fell into this category.

Nearly every disaster described in this book was the result of a mistake. The classic error of all time is, of course, that of the *Titanic.* Since the tragedy has already been covered in many works – I, myself, have written about it in *Stories of Famous Ships* – it is omitted from the present collection. Nevertheless, readers should be reminded that, acting in the belief that she was unsinkable, and ever aware of the sacred schedule which insisted that transatlantic passenger liners should arrive on time, her captain drove her at full speed into an ice field. Indeed, when the *Titanic* approached the ice, she was travelling at 23½ knots – faster than at any other time on the voyage.

But, if the *Titanic* represents the greatest error, she also illustrates the most shameful non-rescue. The drama of her striking the iceberg and foundering just over two hours later took place within full view of another liner. The master and crew of that vessel did nothing; having halted their own ship on account of the ice peril, they dozed through the night – while, about ten miles away, 1,503 people were dying. It was left to the Cunarder *Carpathia* to cover fifty-two miles in the shortest possible time, and to haul the survivors out of the sea.

In *Famous Rescues at Sea,* I have tried to cover as wide a variety as possible. Having written it, my chief feelings are of amazement and awe. Amazement that human beings could endure so much and still live; awe that the combined forces of wind and sea can create such excesses of violence. Perhaps a more seasoned mariner might be less impressed. My own experiences of messing about in boats have usually ended with the boats messing about with me.

It remains to thank the librarians of the National Maritime Museum and the Maidstone Library – also many anonymous newspaper correspondents, especially in *The Times,* who have reported shipwrecks over the past 150 years. Without their help, this book would have been impossible.

Richard Garrett
Tunbridge Wells

1 AN ACT OF GRACE

When the Dundee Packet Company's steamer *Forfarshire* sailed from Hull on her maiden voyage in 1834, she was regarded with awed admiration. Not only was she the finest example of shipbuilding that had been seen for a long time; she was also a dramatic improvement on all previous ways of travelling to the north-west of Scotland. The journey by stage-coach was long, irksome and dreadfully uncomfortable. Now, for those who could afford it, this floating paradise offered all the amenities of an excellent hotel. Indeed, according to her owners, she would extend the very purpose of Dundee. It would become more than a commercial stronghold; it would burgeon into a holiday centre. As they pointed out, the Highlands had so much to offer tourists; the only problem had been that of getting there. Now, with this luxurious steamship providing a regular and reliable service from the south, prosperity could be brought to the hills - with Dundee acting as the gateway.

The *Forfarshire* was, indeed, a most splendid vessel. Her twin 90-horse-power engines were described by a contemporary writer as 'this great propulsive force'. Their power, his rhapsody continued, was 'as high, if not higher, than that of any steamer of the size now afloat'. The steerage accommodation was somewhat basic, but this was not a matter for criticism. Nobody expected the second-class traveller's lot to be anything else. By contrast, the cabin-class quarters were elegant beyond compare. As the same writer enthused, 'Everything is to be done that expense and ingenuity can devise to make them splendid and convenient.'

His superlatives were not misplaced. A well-known artist had been commissioned to decorate the panels in the saloon. There was a marble mantelpiece and thick-pile carpets. A cabin was set aside especially for the ladies; and another for the gentlemen. For the really affluent traveller, a private stateroom could be rented. As for the food, the selection of wines, and the service – they were all impeccable. Nowhere else, on any of the seven seas, were the standards higher.

But this was not all. The *Forfarshire* was *reliable.* Making the journey up the North Sea by sail had been an erratic and sometimes hazardous experience. Everything depended on the wind – the least predictable of the elements, which, with its freakish behaviour, made the time (the day, even) of arrival a matter of luck. With steam engines, it was different. The wind could do what it liked ; these great machines, with their combined output of 180 horse power, would push the ship doggedly onwards. A man knew where he was in the age of steam. He could arrange appointments with a very fair prospect of keeping them.

The *Forfarshire* was so successful that, in 1835, her owners ordered a sister ship. The cost would be £22,000, but there could be no better way of investing money. That, at any rate, was what the directors of the Dundee Packet Company believed. Unfortunately they had overlooked one thing. Despite her power and her elegance, the *Forfarshire* was not flawless. She was certainly beautiful ; she was undeniably fast ; but, apparently unknown to the engineers and the shipping magnates, she had a weak heart. Eventually, this would become apparent and the price would be damnable.

Perhaps it was the result of progress achieved too quickly. The first wisps of smoke had been seen from a steamer's funnel in 1802, when a small vessel named the *Charlotte Dundas* gave a demonstration on the River Clyde. That was a paltry thirty-two years before the completion of the *Forfarshire,* and yet the two ships were vastly different. The *Charlotte Dundas* was little more than a tug ; a craft that was never likely to stray far beyond the mouth of a river. If her engines broke down, she could always be sailed to safety.

The *Forfarshire*, on the other hand, was committed to trade in deep waters. On her passages up and down the North Sea, she was liable to encounter some of the worst weather that any ship is asked to endure. Nor was it consistent. Within a matter of minutes, a gale might suddenly develop ; the once-calm waters became creased up into giant waves, the wind swung in a frenzy from one point of the compass to another. Admittedly, the new packet boat was provided with sails ; but there could easily be occasions when, in the event of an engine breakdown, they would be unable to bring her clear of trouble.

As it happened, the weak link was not this wheezing, clattering assembly of crankshafts and connecting rods, pistons and cylinders, but the boilers that gave them the life force of steam. They were manufactured from iron plates held together by rivets. The method of construction was crude – presently, and nobody knew when, the rivets would work loose and the boilers would begin to leak. If this potential disaster reached its climax during bad weather, the result would be a certain shut-down in the engine room. The *Forfarshire* would be compelled to rely on her sails, which was all very well if the wind were off the land. If, on the other hand, it came from seaward, the chances were that she would be dashed to pieces on the shore.

But at the time such a grim possibility had not been envisaged. Even if it had been, little could have been done about it. There was no alternative method of building boilers ; once they had admitted this flaw in the grand steamship design, the owners would have been compelled to revert to sailing vessels until some ingenious mind dreamed up a better technique. Nor could they postpone sailings to take account of bad weather. One never knew where one was with the North Sea, and there was no way of predicting the conditions. The first official weather forecasts were not produced until 1861 when an admiral named Robert Fitzroy founded the Meteorological Office. In the 1830s, it was impossible to foretell what might be in store for a vessel after she had left harbour.

But, on Wednesday 5 September 1838, Captain John Humble, master of the *Forfarshire*, was not worried by such thoughts. His ship had been in service for the better part of four years, and she had given nobody cause for complaint. During this period, the

North Sea had displayed most of its many and alarmingly various moods, and Captain Humble had never been in trouble. This voyage promised to be like any other – a routine trip from Hull to Dundee, safe, swift and no doubt profitable.

Since passengers paid their fares after they had embarked and the ship had left port, there are no records to show precisely who was aboard. One estimate suggests that the number was sixty-three, of whom twenty-four were officers and crew. Among the twenty believed to be travelling cabin class, there were several businessmen on their way home to Dundee from Russia via one of the Baltic ports and Hull. There was certainly a Mrs Patrick, the wife of another of the Dundee Packet Company's masters ; and a young man named Ritchie who was travelling with his elderly aunt and uncle. The steerage-class passengers included a clergyman, a weaver, a baker, a woman with two children, and a stoker named Donovan who was travelling north to join another of the company's vessels. So far as cargo was concerned, her holds were crammed to capacity with textile machinery, clothing, hardware, cakes of soap, and (ironically) boiler plates.

At 6.30 that evening, the *Forfarshire* cast off from the quay, and headed down the Humber for the sea. By nightfall, she was on course for Dundee, steaming at her normal cruising speed of 9 knots. She had not gone far, however, when the weather worsened, and she had to cut back to 6 knots. By four o'clock on the following morning, she was off Flamborough Head ; the force of the wind was increasing, and the *Forfarshire*'s inherent weakness was at last becoming apparent. The starboard boiler began to leak.

Captain Humble, who had a sailing-ship master's mistrust of technicians, was not unduly perturbed by his chief engineer's reports. Perhaps he thought they were exaggerations ; he certainly gave no thought to any notion that they should put back to Hull. Not least of all, he had the image of his ship to consider. As her popularity with travellers made clear, she had an enviable reputation for speed and reliability. A return to the port of departure would be ignominious. The only people to benefit would be the stage-coach owners.

But the situation in the engine room was getting worse. Later that morning, the boiler room was awash with scalding water, and the stokers were unable to supply the furnaces with coal. The pumps were striving to feed the water back to where it belonged, but the task was too much for them. Two of the fires had to be drawn ; but, in spite of this reduction in the 'great propulsive force', the ship struggled on northwards. Captain Humble, still apparently unconcerned, visited the passengers with one of the stewards. The object was not to reassure them, but to collect their fares. By eleven o'clock that night, they were north of Berwick, somewhere in the region of St Abb's Head.

The night was pitch dark, with unseen storm clouds hurrying across the sky. Violent flurries of rain made the visibility even worse, and the gale-force wind was hurling itself at the ship like a pack of furies. The *Forfarshire* was writhing with a horrible corkscrew motion that seemed intent on maiming her hull. But these were the least of the worries. Down in the engine room, things had deteriorated still more, for the port boiler was now giving trouble. The pumps were completely unable to deal with the water that was gushing outwards in hissing torrents of steam. By an unholy combination of circumstances, the weather was adding to the severity of the crisis. Some time after nightfall, the wind had backed to the north-east. The *Forfarshire* needed every ounce of her strength to punch a way forward ; but the great ship was weak, crippled, with no thrust left in her. At one o'clock in the morning, the chief engineer gave Captain Humble the fatal news. There was no longer enough steam to provide headway. If the *Forfarshire* were to make any progress at all, it would have to be by sail.

Did Humble wish that he had put back to Hull, or into the shelter of South Shields ? Did he curse the pride that had compelled him to continue his voyage against the advice of experts ? There could be no going forward now ; with the north-easter pressing its violence upon the ship, he could only turn about and run before it. The question was - where ? Not many miles to the south-east, there seemed to be the possibility of shelter among the Farne Islands, a scattering of thirty rocky outcrops between two and five miles from the coast. There, surely, they would find safety – an anchorage, where the *Forfarshire* could ride out the storm. Humble ordered the sails to be set and the ship went about.

For the next two hours, she scudded before the gale, struggling to reach the lee of the islands. At some point, a light was seen flashing away to port. It was now raining in torrents ; Humble was uncertain of his position, but it no longer mattered. As the *Forfarshire* was hurled along, it suddenly became clear that something had gone wrong with the steering. Had the rudder been carried away ? Whatever the cause, the ship no longer answered to her helm. She would go wherever the wind and tide wished ; and, at about three o'clock that morning, their intentions became evident. One moment, the look-out was crying 'Breakers ahead !' The next, the pride of the Dundee Packet Company's fleet was being thrown towards a large rock known as the Great Harker. The end was now only seconds away.

Ever since the steering had broken down, the mate had been in no doubt about the outcome of events. The *Forfarshire* was being conveyed towards certain disaster. Nothing could help her now, and he ordered the port quarter boat to be launched. His motive was questionable: the *Tyne Mercury* described his conduct, and that of the eight sailors who went with him, as 'unlike that which is characteristic of British seamen'. On the other hand, a more charitable view suggests that this officer could see the shape of the Great Harker looming up ahead; he knew that the *Forfarshire* was bound to hit it ; and that he felt that, with one of the lifeboats safely in the sea, there would be a better chance of picking up survivors. However, with this deadly combination of wind and current, it is hard to see how such an experienced mariner could believe there was any possibility of remaining in the vicinity of the wreck. And so it was ; hardly had it touched the water, than the craft was swept away from the doomed vessel. It was pushed through a narrow passage between two rocks, and carried out to sea. Some while later, the boat – with all its occupants still alive and well – was picked up by a sloop bound for Montrose. The last person to scramble on board it had been Mr Ritchie, the young gentleman making the trip with his aunt and uncle. As he ran along the steamer's deck, he was seen to be carrying his trousers over his arm. In his hurry to get from his cabin to the boat, there had been no opportunity to put them on. Nor, as he later told a reporter, had there been sufficient time to awaken his relatives. But young Mr Ritchie's evidence was suspect. He claimed that he had jumped five yards from the deck of the ship to the bobbing lifeboat. Even in calm weather, this would have been an impressive feat ; under the circumstances, it would have been impossible.

The first wave lifted the *Forfarshire* up and placed her with comparative gentleness upon the rock. As if this were a mistake, the receding water pulled her away, and allowed the second wave to do the job properly. The stricken ship was lifted up ; and, with all the violence and brutality of the sea in its darkest moments, she was slammed down on to the Great Harker. The impact immediately broke her in two. One moment she was intact ; the next, the entire stern section, with Captain and Mrs Humble, the engineers, most of the crew and all the cabin-class passengers, was sent 'hurrying', as the *Tyne Mercury* put it, 'into another world'. Some of the steerage passengers were carried away, too. One was drowned as he sat curled up inside the cabin, too terrified to save himself. A woman, hanging on to the rigging, somehow managed to clutch on to her two children at the same time. And the waves, not heeding this drama that had been so quickly enacted - with approach, climax, and tragic conclusion all taking place in a matter of seconds - continued their attack on the *Forfarshire's* remains. Captain Humble's last words were, 'My God - we are all lost !' They were true, whichever way you looked at it.

There were now twelve people still alive. Before long, the score would be reduced to nine ; for the two children died, and so did the clergyman who, crouched in a corner of the engine room, meekly surrendered his soul in a state of total exhaustion. There now remained the ship's carpenter, the transient stoker named Donovan, four members of the crew, the baker, the weaver, and the woman whose two children had died. It would not be long before the sea removed the last traces of the *Forfarshire* ; there was no point in remaining on board what was left of her and so, one by one, they gingerly clambered down on to the dripping back of the

Great Harker. For the moment, they were safe ; but the sea might repair this omission in its own good time. There were still two hours to go until high tide. Then, perhaps, it would wipe them, too, away into another world.

About three hundred yards from the Great Harker, there is another rock. It is named Longstone ; it is larger, flatter, a more accommodating place than its grim neighbour. In 1836, a lighthouse had been built upon it. The keeper was a fifty-four-year-old Northumbrian named William Darling. He was the third generation of Darlings to work for Trinity House.

At the age of nineteen, William Darling had married a woman twelve years older than he, Thomasin Horsley. They produced nine children, of whom the youngest was named Grace Horsley Darling. She and her brother, Brooks, were living with their parents at Longstone in 1838. The others had all left home.

Darling was a quiet, deeply religious man. He used to refer to playing cards as 'the Devil's books', and he would not allow any light literature to mar the sanctity of his lonely home. Life for William was a serious business. At first sight, his face appeared to be grim, rather formidable. But presently it became clear that despite his manner, he was a kindly man. Reticence had been built into him by the nature of his work. He was practical ; a man who could handle boats ; a character with a strong sense of duty. No doubt he understood himself ; no doubt he had the same range of emotions as other people. But it was hard to believe, for his withdrawn personality refused to allow him to express either enthusiasm or depression. It was, perhaps, a safeguard in a world where there were too few people on the one hand - and, on the other, where those who *were* around him lived in too close proximity. Something had to be done to compensate for this ; a man had to erect defences even against those he loved. Effusiveness of any kind would be intolerable ; utterances of affection could just as well become declarations of hatred. It was best to keep things neutral.

When the *Forfarshire* was wrecked, Grace Darling was twenty-one. She was on the short side - five feet two inches tall -

but well proportioned and with a very pleasant personality. She had fair hair and soft features. Although a visitor to the lighthouse once described her as 'perfectly free from the shy, awkward gait of country girls in general', she had some of her father's reserve, and another source suggests that she blushed easily.

Mrs Darling must have been a capable housewife, for she had succeeded in transforming the austere interior of the lighthouse into a comfortable home. As for Brooks, he had inherited the Darling flair for good seamanship, and he spent much of his time fishing. On the night of 6 September he was away from home ; the herring season had begun, and he was staying five miles away at Seahouses.

At about the time the *Forfarshire* was hurled on to the Great Harker, William Darling was minding the light ; Grace and her mother were asleep. But, as he watched the gale gathering renewed fury, and he knew that an unusually high spring tide was expected, he became worried about their fishing cobble. It was too close to the water's edge ; there was a danger that it might be carried away by the sea. Normally, Brooks would have helped him to move it, but his son was not there. Grace would have to give him a hand. He went down to her room and knocked on the door.

She dressed hurriedly and, between them, they managed to haul the boat to safety. With the wind screaming past the rock, the work took some time, and it was nearly three o'clock before they were able to go indoors again. Mrs Darling was now minding the lamp ; her husband retired to bed and Grace went back to her room.

Although she was very tired, she could not sleep. At about 4.45, she climbed out of bed and walked across to the window. A pale watery dawn was creeping into the sky. Ahead of her, she could make out the gaunt black shape of Great Harker. She had seen it so many times, but now there was something different about it. She peered into the murk, straining her eyes to see what it was. Then, with a sudden shock, she realized the truth. Lying on the hump of the rock, shrouded from time to time by the crash of waves, lay the remains of a large steamship. As William Darling reported to

Trinity House afterwards in his statement about the tragedy:

> One quarter before five, My Daughter observed a Vessel on the Harker's rock, but owing to the Darkness and spray going over her, could not observe any person on the Wreck, although the glass was incessantly applied until near seven o'clock.

When she saw the remains of the *Forfarshire*, Grace went immediately up the staircase to the lantern, and told her mother about the awesome spectacle. Reluctantly, Mrs Darling was persuaded to rouse her husband. Between them, they kept the Great Harker under continual scrutiny until, at last, it was light enough to make out some of the details. Then, shortly after seven, they saw a movement. Presently, there were other signs of life. Three – or was it four ? – people could be discerned. The gale was blowing as fiercely as ever ; somehow, these people must be brought to safety. With Brooks Darling away from home, such a task seemed to be impossible. By now, the lifeboat at Seahouses would certainly have been alerted by the look-out at Bamburgh Castle, but it seemed unlikely that it could make any headway against the onslaught of the north-easter. No – the inhabitants of Longstone were the only hope. A man aged fifty-four, an elderly woman, and their comparatively frail daughter – that, too, was impossible. Mrs Darling would have to remain behind ; but how could her husband and Grace cross that deadly three hundred yards of seething water? Mr Darling was by no means convinced. Grace, on the other hand, had already made up her mind.

> I had little thought of anything but to exert myself to the utmost, my spirit was worked up by the sight of such a dreadful affair that I can imagine I still see the sea spray flying over the vessel . . . It was sufficient to affect the strongest nerve to view the wreck.

In rough weather, it took three men to manage a fishing cobble. But William Darwin could see that it would be possible to use the wind and the tide to carry them into the shelter of the Harker rocks. After that, the going would be comparatively easy. The hard part would be making the return trip, but there seemed to be a number of men in the vicinity of the wreck. If these were sailors, they should be able to manage it.

Mr Darling was still doubtful, but Grace was determined. If he refused to go, she would make the journey on her own. It would mean almost certain death ; a brave act of defiance probably accomplishing nothing. But this was better than the prospect of a life made hideous by the mute reproaches of those drenched and forlorn figures she had seen clinging on to the rock. Eventually William Darling realized that there was nothing else for it ; he and Grace would have to make the attempt.

Outside the snug interior of the lighthouse, it was pouring with rain. Grace had dressed for the occasion by removing all her petticoats, and wrapping a plaid shawl tightly around her neck. She also took with her some blankets. The survivors had been on the rock for several hours. Some might be injured, others suffering from exposure. It seemed likely that they would be needed.

The distance between Longstone and the Harkers could be measured in yards. But, by the route taken by the Darlings, it was over a mile. First there was the headlong rush, as the wind took over the craft – hurling her forward on the tops of the giant combers that rolled in from the ocean. When, at last, they reached the shelter of the reef, the going was still difficult. A heavy swell was running ; violent eddies grasped the small craft and spun her round ; now and again, the top of a partially submerged rock threatened to smash the bottom out of her.

As they approached the Great Harker, they could clearly see the remains of the steamer. She was bigger than they had imagined ; lying on her side and, where the stern section had been swept away, they could see the shattered cabins. The wave-battered, spray-drenched survivors were clinging on to a spikey ledge. There seemed to be about a dozen of them. It was more than could be accommodated in the boat which was only twenty-one feet long by five feet wide. And the hardest part was yet to come.

When they saw Grace and her father approaching, the castaways hurried to the edge. A woman was holding on to the corpses of her two children – who had now been dead for some hours. A hunched-up figure in the lee of the *Forfarshire*'s hull turned out to be the clergyman's body. Those who were still alive began to cry out. Panic had taken hold of them, and William Darling assessed the situation in that practical way of his. Somebody had to assert leadership ; to bring them to their senses and to shepherd them calmly into the boat. Otherwise, they would rush it. The boat would be swamped, and they would all be killed in the vicious sea that boiled around the rock.

Grace would have to take charge of the craft. He would climb on to the Great Harker's back. To leave a young woman in charge of a twenty-foot boat ; to abandon her to a sea that one moment seemed intent on smashing her against the rock, the next appeared determined to snatch her away for ever ; this, surely, was asking a great deal. But Darling had no alternative. As he admitted afterwards, this was one of the more horrible moments of his life. But, with these fear-crazed survivors, what else was there to do ?

William Darling clambered on to the rock. He was, as always, quiet. But he never shouted ; there was never any need. This strange, withdrawn man had strength and determination – qualities that, by some miracle, communicated themselves to the survivors. The ship's carpenter, Thomas Tickett from Hull, and two other seamen agreed to relieve Grace. On the first trip, they would take the woman and a badly injured man. After that, they would return for the remainder.

But there were still problems. The woman refused to be parted from the bodies of her children. It had to be done by force ; fortunately, she was too weak to put up much resistance. Throughout this period, Grace had managed to keep the boat under control. Somehow, they managed to lower the woman and the sick man on board. After that, it was comparatively easy. With the *Forfarshire*'s sailors at the oars, they made short shrift of the journey back to Longstone. As William Darling recalled:

> There were five seats, one of them astern. Grace sat on the mid-ships, the woman sat upon the flooring forward with her head lying against the side, as she was not fit to sit on the seats

and one of the men was lying aft in the same manner with a blanket around each. The children were both left in the wreck, dead. Four oars used in pulling back, no rudder shipped.*

Once they reached Longstone, Grace, the woman, and the sick man made their way to the lighthouse ; the others went back to the wreck to bring off the others. It was still blowing hard. As William Darling rightly predicted, it would be three days before any boat could make the journey from the mainland. But, not long after the rescue had been completed, another craft arrived at Longstone. Somehow, Brooks Darling and seven members of the North Shields lifeboat crew had done the impossible. They had managed to launch a boat from the beach ; they had rowed it five miles against the battering force of the gale, and they had reached the Great Harker. But, by the time they arrived, William Darling had completed his work. Now, they too were forced to seek shelter in the lighthouse, for they were unable to make the trip back to Seahouses. At some point in a journey that had taken them two and a half hours, they had noticed the steamer *Liverpool* bound north from London. They had signalled her for assistance ; but the captain – wisely, perhaps – had kept on his course and ignored the plea. The corpse of one vessel on the Great Harker was enough for one day.

The first news of the disaster and its heroic sequel was given to the press by the driver of the *Royal William* – a coach bound for Dundee that had carried some of the survivors. People in the vicinity, it seemed, had been generous. Within two hours of the plight of the survivors being known, the sum of £11.40 had been raised by public subscription. As for the stage-coach operator, he had been especially munificent. He offered to carry the stranded passengers to Dundee free of charge – and to provide them with hotel accommodation on the way. But perhaps he could afford to be good natured. The fate of the *Forfarshire*, and the enormous impact it made in the public prints, was bound to benefit his business.

* Letter to an artist who was making a painting of the rescue.

On 4 October, the directors of the Glasgow Humane Society sent Grace Darling their honorary silver medal 'in admiration of her dauntless and heroic conduct in saving (along with her father) the lives of nine persons from the wreck of the *Forfarshire* steamer, September 6th, 1838'. Grace replied, and her letter was printed in *The Times.*

> Be assured that I lament most sincerely the awful loss of human life caused by the wreck . . . It affords me great pleasure to think that my humble endeavour, assisted by Divine Providence, has been instrumental in saving the lives of nine persons ; but ah ! how much my heart yearns within me to think how many valuable lives have been lost to their mourning friends, and to society at large, on this most melancholy occasion.

All of Britain except that section living in the area of the Farne Islands raised Grace Darling to the status of heroine. Woman wrote to her, asking for locks of her hair. Artists lined up to paint her portrait. Public subscriptions were raised on her behalf, and so it went on. But, in this corner of Northumberland, nobody thought much about it. Grace herself was modest about her feat, her father never mentioned her assistance in his official reports to Trinity House. As for the local population, they either refused to answer questions put to them by visitors, or else they tried to underrate the achievement. According to some of the stories they told, it had taken place at low tide, and a boat had not been used at all. The survivors had, simply, walked to Longstone. It was not that they had anything against Grace ; just that, like the Darlings, they could not see anything extraordinary about it. Too many ships had been wrecked on the Farne Islands to make the occasion remarkable.

The last grisly vestige of the *Forfarshire* tragedy was taken from the sea by a sailor in the steamer *Superb* on passage north from South Shields. It was the body of a woman dressed in a morning gown. She was wearing a gold watch and two rings. She also had with her a sovereign, 37½p in silver, and a bunch of keys. Her linen was marked 'L.M.P.', but one of her boots was more informative. Inside, it was labelled 'Mrs Patrick'. She was the wife of a Dundee Packet Company's captain, who had been on her way to join her husband in Scotland.

2 THE COURAGE OF CORNELIUS BRADY

At about six o'clock on the morning of 1 April 1873, the villagers of Prospect, Nova Scotia, were awakened by a dishevelled young man wearing the uniform of a third officer in the White Star Line. His name, it seemed, was Cornelius Brady ; he was serving in the pride of his company's fleet, the Royal Mail Steamship *Atlantic*. To judge by his brief account of the past three hours, young Mr Brady had been having a terrible time. His ship had struck a rock offshore. Many people were dead ; unless help came quickly, the casualty list would be a great deal longer.

Cornelius Brady remained at Prospect until noon that day. Then, having seen that the rescue operations were going as smoothly as the weather allowed, he set off for Halifax. He arrived in the city at midnight, and the second part of his mission began.

As he was only too well aware, the first consideration in a shipwreck is to bring the survivors ashore. This may be the complete solution to the problem; but, when the amenities on land are no more than a thinly populated fishing community can provide, and when there may be as many as four hundred people involved, there is more to it. Somehow, these stranded travellers have to be conveyed to more civilized surroundings ; the question, in this instance, was – how ? There was no rail link between Prospect and Halifax ; and even Mr Brady, who was in a hurry, took a long time to make the journey. There was, surely, a better way than travelling overland.

During these eventful hours, the young officer was doing more than his reasonable share of ripping people unceremoniously from sleep. First it had been the fishermen of Prospect ; now it was the officials of the shipping companies in Halifax. He must have moved quickly and effectively ; for, three hours later, the Cunarder *Delta*, the Dominion Line's *Lady Head*, and the tug *Goliath* were all preparing to set off for the scene of the disaster. He had done as much as anybody could do. By no means dissatisfied with his performance, Cornelius Brady took himself off to an hotel, a hot meal, and a comfortable bed. Those who had survived the sinking of the *Atlantic* were, he felt sure, in good hands.

The liner *Atlantic* was a vessel of 3,707 tons. She had been built two years previously by Harland and Wolff of Belfast at a cost of £120,000. In addition to her 600-horse-power Maudesley steam engines, she had four masts - three of which were square-rigged. But this concession to the now vanishing days of sail was little more than a convention. The *Atlantic* relied entirely on her machinery. If it broke down, or if she ran out of coal, she was in trouble.

On 20 March, she sailed from Liverpool for New York. At eight o'clock on the following morning, she called at the Irish port of Queenstown (now Cobh), where another 200 passengers embarked. When, at 12.30 that day, she turned towards the Atlantic, she had 1,038 people on board, 1,836 tons of assorted cargo, and (this was to be important) 967 tons of coal.

Throughout the crossing, the weather was bad. After eleven days at sea, the liner was somewhere off the coast of Nova Scotia. New York was still more than 460 miles away ; a storm was forecast ; and the master, Captain James Williams, was becoming worried about his supplies of coal. He ordered the speed to be reduced to 7 knots and gave the helmsman a change of course. He proposed to put into Halifax to refuel.

Was this action necessary? The *Atlantic*'s consumption of coal over an average of eighteen voyages across the North Atlantic had been 744 tons a trip. The most she had ever used up had been 896 tons, and this had been in mid-winter when conditions were far worse than on the present passage. With 967 tons initially in her bunkers, there should have been more than sufficient. Indeed, according to one estimate, she still had enough in hand for five days. Since the last leg of her voyage to New York would take, at the most, forty-eight hours, there cannot have been any great cause for concern.

Or had an excessive amount of fuel been used up by driving the vessel too fast? The eleven-day trudge across the ocean does not suggest that Captain Williams was bent on breaking any records ; nor did company regulations permit such feats. 'No supposed gain of expedition, or saving of time on the voyage,' they stressed, 'is to be purchased at the risk of an accident.' And, in a letter to Williams on his appointment to the *Atlantic,* the owners wrote:

> We invite you to dismiss from your mind all idea of competitive passages with other vessels, the advantage of success in which is merely transient, concentrating your whole attention upon a cautious, prudent, and ever watchful system of navigation, which shall lose time or suffer any other temporary inconveniences rather than run the slightest risk that can be avoided.'

Why, then, did he decide to make for Halifax ?

Later, however, his actions became even more mysterious. Williams, it should be made clear, was a man of forty-three with an excellent record. Before joining the White Star Line when the company inaugurated its transatlantic service in 1870, he had served as commodore to the Guia Line. He held an extra master's certificate ; and, according to one of the passengers, 'he never drank anything stronger than lemonade'. This was his second voyage in the *Atlantic* ; the previous trip had been without incident, but he had never before been to Halifax.

Here, then, was this model master mariner approaching an unknown coast, which was reputed to be dangerous, at dead of night. Nevertheless, at midnight, he ordered an increase in speed to 12 knots (since they were now so close to Halifax, there was no more need to economize on fuel). The ship was to remain on her present course until 3 am, when he would stop engines and wait for daylight. Having given these instructions, he handed over to the second officer, and retired to the chart room, where he gave an interview to a feature writer from the American newspaper *Cosmopolitan*. The man was, it seemed, writing an article about the ship. The conversation lasted twenty minutes. After that, Captain Williams went to bed and was soon fast asleep.

He was, presumably, a deep sleeper. At all events, when his steward visited his cabin at 2.15 to rouse him, he refused to stir. Before joining White Star, this member of the crew had served for several years with Cunard. Afterwards, he said that, in all his long experience, this was the first time he had ever known a captain sleep while his vessel was approaching a hazardous coast.

No doubt, if he had been aware of the true situation, Williams would have slept less soundly. For, by some error of navigation, the *Atlantic* was off course. When the second officer mistook one lightship for another, the error became even more serious. Unless some miracle occurred, she was bound to strike one of the rocks off Mars Island. But there were no miracles that night. When the news of the disaster was reported in the press, a correspondent of the *Liverpool Daily Albion* observed that, 'Read by the light of the catastrophe, it almost seems as if they had felt a sort of presentiment of some possible future evil.' If Captain Williams experienced any such emotion, he did not allow it to disturb his slumbers ; nor was anybody else on board busy foretelling disaster. When the ship struck, most of them were tucked up in their bunks. Nevertheless, the remark was not entirely fanciful. There are so many factors that were never satisfactorily explained, that it almost seems as if the *Atlantic* and her human freight were the victims of destiny.

One passenger who was not tucked up in his bunk was a gentleman named Hogan. Mr Hogan had been unable to sleep and, just before three o'clock, he decided to take a turn on deck. He heard the look-out call 'All's well! Three o'clock', and then about ten minutes went by. By this time, according to Captain Williams's plan, the vessel should have been hove-to. But no such orders were given. She steamed steadily onward until, suddenly, Mr Hogan heard one of the sailors exclaim, 'Breakers ahead!' About three hundred yards away, according to his reckoning, he saw a rock. 'I thought at first it was an iceberg,' he recalled. 'Then the ship commenced to heave and plunge, and after three or four plunges, she settled and the water seemed to be rushing in.' He decided that, in what was obviously an extreme emergency, the landward side of the vessel would be safest. He ran across the deck and caught hold of the rigging as, slowly at first but then gathering speed, the *Atlantic* began to turn on her side.

'I then heard a dismal wail which was fearful to listen to,' Mr Hogan's story continued. 'It proceeded from the steerage passengers who were below, and who were then smothering. It did not last more than two minutes, after which all was still as death.' The next thing he heard was Captain Williams's voice, crying out that the only chances of survival were to climb into the rigging.

A passenger in one of the cabins was awakened by the impact. 'Almost instantly,' he said, 'I heard a tremendous crash, and the air rushed in and blew out the lights. It was my impression that the boiler had burst.' At first, he made no effort to save himself. 'I said it mattered little – thinking the steamer was out at sea ; and I said that it was just as well to meet death below as on deck.' Eventually, he bestirred himself and went on deck. But there were many who died in their bunks. In the saloon, an elderly couple named Mr and Mrs Fisher were standing near the door. Mrs Fisher was heard to say, 'Go – save yourself.' Her husband replied, 'No, we will remain together and meet in a better world.' They held hands, and sat down on a settee.

Such dignity in the face of death was exceptional. One old man offered $100 to anybody who would save his life. There were no takers. Others made a concerted rush for the companionways. Many were knocked down, trampled on and died of suffocation. When, some days afterwards, divers visited the wreck, they found that almost as many had perished by this method as by drowning.

The vessel was obviously doomed, and it was now equally clear that her death throes would not last long. The propeller had struck the rock with such force, that the impact started the rivets and wrenched the stern apart. Immediately, water poured into the women's compartment of the steerage accommodation, drowning all the occupants. Then, as the *Atlantic* rolled over on to her side, she snapped into two pieces just aft of the foremast. The boilers burst ; soon, only the masts and a section of the starboard flank of the hull remained above water.

During these final moments, the situation on deck was little short of chaotic. Women and children were screaming ; men were shouting ; the crew were urging the passengers to climb into the rigging ; and, at one-minute intervals, this bedlam was punctuated by the sound of a distress rocket soaring into the sky. Would anybody see them ? Would anyone come to their help ? Nobody now knew precisely where they were. The coast, which could not be very far away, might be uninhabited.

Moments after the *Atlantic* hit the rock, Captain Williams ordered one of the boats on the port side to be lowered. It was crammed with about forty people. Almost simultaneously, however, the ship keeled over. The craft was thrown into the sea and all those on board were drowned. Immediately afterwards, the remaining lifeboats on the port side were washed away; and, once the liner had settled, it became impossible to lower those on the starboard side.

The wind, the rough sea, and the cold were inflicting terrible punishment. Captain Williams helped two women into the rigging. Then he walked across the deck to supervise the lowering of the ill-fated lifeboat. When he returned, they had been washed away. Another woman was lashed to one of the stays. 'But,' said a survivor, 'when I saw her from the shore, her body hung downwards. I think she must have frozen to death.' Of the 401 people who came through the ordeal, there were no women and no children.

Some people, including Captain Williams and the chief officer, Mr Firth, sought sanctuary in the rigging. Others crouched on the side of the hull that remained above water. A few more managed to climb down on the rock. But there was a limit to the length of time they could remain in any of these places. They would either be washed away by the rising tide or die of exposure. The ship's purser was one of those who froze to death while clutching at a rope. Several of those on the rock suffered the same fate. Shouting from his own perilous perch, Captain Williams urged them to keep moving. Many of them now experienced a feeling of lassitude that strangled all further initiative. It was tempting to lie down and sleep, but this would be fatal. Above all things, Williams exhorted them, remain awake.

The rock on which the *Atlantic* had foundered was about forty yards from a much larger outcrop named Mars Island - which, itself, was a matter of 150 yards from the shore. Third Officer Brady was an unusually strong swimmer - and so were two of the ship's quartermasters, a pair of robust seamen named Owens and Speakman. Between them, they decided it would be possible to get a rope across to Mars Island. After that, they might reasonably hope to carry a second lifeline to the mainland. Mr Brady tied one of the halyards round his waist, plunged into the rough sea and struck out for the first objective. Against all the laws of probability, he reached it. With the help of Owens and Speakman, he managed to get three more lines across. All told, fifty people reached safety by this method, though many (among them the chief steward) were drowned in the attempt.

Mr Brady was not a man to hang around. Having completed the first part of his task, he set off on the second stage of his dangerous journey. At last, cold, wet, and by now very tired, he reached a habitation named Prospect. It was nearly six o'clock on what promised to be a fine day. From the shore, it was possible to make out the remains of the *Atlantic* with her pitiful complement of survivors. The local fishermen acted quickly. Within minutes of Mr Brady's arrival, a boat had set off for the wreck. The rough sea made rescue operations difficult, but there was encouragement from two sides.

Captain Williams called down an offer of $500 for every boatload of humanity brought safely to shore. Two other fishing boats were launched; and, on land, an ingenious citizen of Prospect removed the blackboard from the village school-room. On it he chalked the message CHEER UP, THE BOATS ARE COMING TO YOUR ASSISTANCE. It was greeted by a surprisingly loud outburst of cheering.

The pandemonium on board the *Atlantic* had subsided some while ago. People were either too tired, or too resigned, or - now that there seemed to be hopes of rescue - sufficiently optimistic to recapture their dignity. When Mr Hogan was taken off by one of the fishermen, he estimated that there were eighty people left in the rigging and on the side of the ship. He described their attitude as 'cool and confident of rescue'. When Captain Williams was taken ashore, there were only fifteen people left. The master would, of course, have been the last to leave. But his place was becoming threatened by the rising tide. If he had tried to remain there any longer, he would have been swept away.

Later in the morning, the storm that had been one of the reasons for the *Atlantic*'s change of course to Halifax, seemed to be materializing. By noon, it had become too rough to make any more trips to the wreck ; but four hundred people had been saved – only one man remained behind. This was the luckless chief officer, Mr Firth, who was hanging on to the shrouds of the mizzen-mast. At first it seemed that Mr Firth was being abandoned to certain death. But, at three o'clock that afternoon, there was a lull in the weather. A boat went out, and he was brought alive and triumphant to shore.

The inhabitants of Prospect were nothing if not hospitable. The survivors were accommodated in the fishermen's huts. They were given food ; huge fires were built to dry them out ; now, at last, they could sleep. Most of them were bruised, some were more seriously injured. Their clothes were in tatters ; and none of them had been able to bring any belongings ashore. But they were alive, which was more than could be said of two-thirds of those who had sailed in the *Atlantic* on this voyage.

Meanwhile, the busy Mr Brady had been attending to the final part of his mission. By noon on the following day, the liner *Delta,* the *Lady Head,* and the tug *Goliath* had arrived on the scene. Three hundred people were embarked in the *Delta* ; seventy-one in the *Lady Head.* Others had already followed Mr Brady's example, and set off to walk to Halifax. An officer and four seamen were left behind to guard the wreck and the bodies that were now drifting in from the sea.

The coffin-makers and hospitals of Halifax were alerted for what were clearly going to be a busy few days. The Nova Scotia Legislature, which was in session, interrupted its business to vote money for the care of survivors and for the burial of the dead. After a short stopover in Halifax, the bedraggled assembly of refugees, many of them now wearing clothes given to them by the fishermen of Prospect, were taken by ship to Portland, Maine. From there, they travelled to New York by train. While they were still in transit, the New York Wreckage Company was instructed to begin salvage operations on the *Atlantic.*

It would, the directors of the White Star Line estimated, cost £150,000 to build a replacement for the lost liner. Naturally, the vessel was insured - but since the company was, to some degree, acting as its own underwriter, this was no great comfort.

When news of the *Atlantic*'s disastrous end reached the ears of the press, it created a furore. The *Western Morning News* described it as, 'The most terrible wreck that the world, with one exception, has ever known ; and, for that exception, we have to go back nearly two centuries.' The newspaper went on to list some of the major ship disasters that had taken place in the nineteenth century. There was HMS *Captain,* which had capsized in the Bay of Biscay, with the loss of 500 naval officers and ratings ; the emigrant ship *Royal Charter,* which went aground off Anglesey with the loss of 446 ; and the troopship *Birkenhead,* which struck a rock off the south-eastern coast of Africa - with a toll of 454. The *Atlantic*'s score of 615 dead (448 men and 167 women) put all these statistics into the shade. It could only be compared, the *Western Morning News* told its readers, with the fate of Admiral Sir Cloudesley Shovel's flagship *Association* - which, in 1707, had gone down off the Scilly Islands with the loss of all the 800 men aboard her.

The White Star Line had received detailed news of the disaster shortly after the *Delta* and the *Lady Head* had begun embarking survivors. It was conveyed in a cable that read: 'Passengers expected at Halifax this evening. Cunards at Halifax are caring for them. Captain Williams mistook the Sambro for the Devil light. Struck ledge two on Tuesday morning. Rolled into deep water almost immediately. No women or children saved.' Somehow, the New York press got hold of the text and proceeded to prejudge

Captain Williams. One newspaper blamed the catastrophe on 'the behaviour, incompetency, and neglect of the captain'. But the shipping line did not escape unscathed. As soon as the reason for the *Atlantic*'s diversion to Halifax became known, it was assumed that she had set off from Liverpool with too little coal in her bunkers. This, the same newspaper exclaimed, was due 'to the criminal parsimony of the company'.

At the same time as these views were published, the New York offices of the White Star Line were being besieged by anxious and excited crowds. An ominous atmosphere became even worse, when rumours reached the city that the *Atlantic*'s crew had behaved in a shameful manner. According to one account, they had conducted themselves insubordinately throughout the voyage. Another asserted that they had been guilty of mutilating and robbing corpses. The former was said to be 'an exaggeration' ; the latter was not the work of White Star sailors at all, but of eight stowaways who had travelled in the ship. All of them escaped from the wreck ; and, before they were brought into captivity, they had been busy pillaging the dead.

Eventually, a court of inquiry was convened to determine the precise causes of the disaster. Much of the evidence centred around the question of coal. Had the liner carried sufficient ? The board of trade surveyor of steamships for the Port of London expressed the view that the amount should have been adequate – unless, he said, the chief engineer had been guilty of 'careless driving'. Since this official was a former White Star employee, his views were suspected of bias. The emigration officer at Liverpool, whose duty was to ensure that vessels put to sea with adequate quantities of all the necessities on board, was also questioned. Yes – he had been satisfied. But the gist of his statement was that there had been more food in her storerooms than the Passenger Act of 1855 required. This was no doubt interesting, but it did not shed very much light on the question of fuel. In the end, the court decided that the White Star Line should have supplied more.

But the public's interest centred more upon the conduct of the captain. As a glance at an atlas showed, he had been disastrously off course. He had assumed that he was approaching Halifax harbour when, in fact, he was heading for a point on the coastline about thirty miles to the west. Captain Williams pointed out that, throughout the voyage across the ocean, there had been a series of gales - some of them from the west, some from the east. Furthermore, and he called a fellow master mariner to confirm it, there had been an unsuspected current, between fifty and sixty miles wide, running along the coast of Nova Scotia at 4 knots. This could well have accounted for the error ; and the confusion about the two lights had done the rest.

Mistakes were easy to make and, sometimes, they could be excused. But this did not explain why Captain Williams had gone to bed - leaving the conning of the ship to one of his subordinates. As *The Times* not unreasonably pointed out, 'If a first officer is to take charge in cases of emergency, the captain might as well be dispensed with.' The court took a similar view, though it treated Williams with leniency. It found that he was, indeed, guilty of neglect by going to bed. Normally, this would have caused his master's certificate to be cancelled. But, the members decided, once the *Atlantic* had hit the rock, he had extended himself in his efforts to save life. For this reason, they were merciful ; it was merely suspended for two years.

Meanwhile, divers had managed to get inside the wreck. They discovered that the back of the liner had been broken at some stage. By a stroke of good fortune, the cargo was still intact ; but, having made this discovery, their attention turned to a more grim aspect of the disaster. At one end of the steerage accommodation, they discovered the bodies of 226 passengers who had been struggling to reach the deck. For the rest of his life, Captain Williams would have to live with their faces in his imagination. Beside this ordeal, the suspension of a certificate of competence for two years was nothing.

Indeed, the only person who came away from the wreck of the *Atlantic* with any credit was Third Officer Brady. If he had not taken that line ashore, the casualty list would have been very much greater. He was the hero of the occasion - everyone else, in one way or another, was a victim.

3 THE DEAD OF THE NIGHT

Eighteen hundred and seventy-five was a sad year for German shipping. Within a period of seven months, two of the country's crack transatlantic liners were destroyed by the sea; hundreds of lives were lost; and, in each case, the disaster occurred close to the very thing that should have given a warning. In the first instance, it was a lighthouse; in the second, a lightship.

So far as the keepers of the Bishop Rock lighthouse, four and a half miles south-west of the Isles of Scilly, were concerned, the night of 7 May 1875 was just like many another. There was, admittedly, dense fog, but there was nothing uncommon about that. At some time, round about ten o'clock, one of the men thought he heard a ship's gun, Again, it was not remarkable. Vessels travelling from America to Britain frequently discharged 'report' guns to announce the completion of a safe crossing. It was certainly not enough to suggest that, within a mile of the lighthouse, the pride of Germany's mercantile marine was breaking up – and that, of the 372 people on board, 329 were drowning.

On the following morning, they became aware of what had occurred; but, by then, it was too late.

Obadiah Hicks, a boatman who lived on the island named St Agnes, had also heard the gun. He put the time at somewhere between midnight and 1 a.m. To his way of thinking, it could only signify one thing: somewhere out to sea, a ship was in distress. He walked down to the shore, but he quickly realized that he was unable to do anything about it. The sea was too rough; and, even if he did go out in his boat, the fog would destroy any hope of finding the stricken vessel. Whatever action he decided to take would have to wait until daybreak.

At four o'clock that morning, Mr Hicks resolved to try again. He took six men with him. For the next two hours, battling against heavy seas, they searched the waters round about. They saw nothing. At six o'clock, they were about to return to St Agnes when, suddenly, the fog lifted. There, less than a mile away, they saw the masts of a ship. It was obviously in trouble ; the fore-mast was missing and there was no sign of the hull. Then, they heard shouting. When, at last, Mr Hicks and his men returned to the island, they had seven survivors with them. One of them had been picked out of the sea, lashed to a wooden plank.

The St Mary's lifeboat was alerted. Towed by the steamship *Lady of the Isles*, she set course for the wreck. Other boats from the Scillies joined in the search ; between them, they picked up forty-three people. The rescue operation, such as it was, was over.

The *Schiller*, a screw-driven steamship of 3,421 tons, had been built by a Glasgow shipyard two years previously. She was the finest and fastest vessel in the service of her owners, the Adler Line of Hamburg. Her master, Captain Thomas, was an Englishman who formerly had been employed by the P & O Line. He was considered to be exceptionally good at his job. Apart from his substantial reputation as a seaman, he was popular with his crew and with the passengers. One of the former described him as 'the favourite officer of the fleet'.

On 27 April 1875 the *Schiller* sailed from New York bound for Plymouth and Hamburg. Captain Thomas had orders to make a landfall at the Isles of Scilly, where the coastguard would telegraph his approach to the Adler Line's Plymouth office. There were 254 passengers on board and 118 crew – plus a mixed cargo, 253 sacks of mail, and $40,000 in gold coins.

The voyage was fraught with difficulties. On the way out of New York harbour, the liner's propeller was fouled by a fishing net. This delayed the departure ; and, when at last she got under way, the *Schiller* soon sailed into bad weather. The last three days leading up to the night of 7 May were particularly bad. Strong winds made the motion of the ship so uncomfortable, that few of the passengers were present at meals. Overhead, the thick covering of cloud made it impossible to take any observations. Captain Thomas had to rely on dead-reckoning – with the result that, when fog came down on the afternoon of the 7th, he was some way off his course. When, finally, the ship struck the Retarrier Ledge near Bishop Rock, she was eight miles from Thomas's estimated position.

During the day of the 7th, the sea moderated slightly ; there were more passengers to be seen about the ship ; and when, early in the evening, one of the officers celebrated his birthday, a fair number of people attended the party.

Throughout the voyage, *Schiller* had been making a steady 14 knots. When visibility dropped, Captain Thomas reduced the speed to 7 and doubled the lookouts. The liner groped on towards the Scillies, blinded by fog and battered by waves until, at about 10 pm, she struck the rock. Strangely, the impact was so gentle, that only a few people were aware that anything out of the ordinary had happened. There were, indeed, rumours that the vessel was approaching land ; some of the male passengers went on deck, expecting to see the lights of the Cornish coast.

About an hour earlier, the wind had increased. Great waves were now seizing hold of the liner, slamming her up and down on the rock and battering holes in her hull. Far from lifting, the fog seemed to be becoming thicker than ever. According to one of the survivors, it was impossible to see one end of the ship from the other.

There was obviously no point in sending up rockets on such a night ; nobody would see them. Consequently, Captain Thomas did the next best thing – he ordered the ship's gun to be fired until the powder became so wet that it was impossible to use it. This was the sound heard by the lighthouse keepers and by Obadiah Hicks (who seems to have got the time wrong).

When the plight of the vessel became known, panic took hold of the women passengers. Some of them clung to the rigging ; others clung to Captain Thomas. At one point, their terror became so great, that some sort of a stampede seemed to be imminent. Inevitably, a number of people would have been trampled to death ; in an effort to restore order, Thomas fired his revolver over their heads. One of the survivors thought this was completely justified. 'I think that if I had been in his place,' this gentleman said, 'I should have shot *at* them.' He was still trying to calm this shrieking mass of womanhood when, at 2 am, the bridge was carried away. Captain Thomas, the chief engineer, and a number of passengers went with it. The Captain's last words were to the second officer who was watching helplessly from the deck below. 'Goodbye, old fellow,' he called. 'Remember me to my friends.' It was, a passenger said, typical. 'A more clever gentleman, and a better sailor than the Captain,' he said, 'there could not be.' The same witness described Captain Thomas's conduct throughout as 'cool and cheerful'.

Thomas was a brave man, there can be no doubt about it. But the state of mind of the panic-stricken passengers is understandable. Giant seas were hurling themselves at the ship, which was steadily falling to pieces. The force of the wind was like a battering ram, mindlessly, brutally assaulting people who had been snatched from sleep and the warmth of their cabins with the sudden realization that, at any moment, they might die. As if this were not terror enough, the scene was shrouded in a thick wrapping of fog. They could hardly see, and they could certainly not be seen. In their agony, they were encapsulated, hopeless, beyond the possibility of rescue. Within this cocoon of hell, in the middle of nobody knew precisely where, they were at the mercy of elements that understood nothing and could show no mercy. The eeriness of it, the despair, was enough to drive anyone out of his mind.

When it became clear that the heavy seas were intent on demolishing the *Schiller*, the crew made several attempts to launch the lifeboats. Only three of the eight on board got away. The others were smashed and their remains carried away like so much drift wood. Several people lashed themselves to the fore mast. When it collapsed, they were trapped. Ironically when, later, bodies were washed ashore on the Isles of Scilly, they were all wearing the so-called 'unsinkable life-jackets'.

It was, perhaps, surprising that anybody survived that ordeal. Even if Obadiah Hicks had been able to reach the ship when he heard the guns, it is doubtful whether he could have achieved much. The number of people at risk, the size of the waves, the speed at which the *Schiller* was breaking up – they all added up to an impossible task.

All that survived from the *Schiller* were forty-three human beings, twenty-three sacks of mail rescued by the St Mary's lifeboat, and the scattered remains of cargo that were smashed up on the Isles of Scilly – no doubt to the benefit of the islanders. When, later that year, the second German shipping disaster of 1875 occurred, well over fifty people perished. Afterwards, pointed questions were asked. Why, the newspapers wanted to know, had help been so long in arriving? For thirty-six hours, the crew and passengers of a transatlantic liner were at the mercy of the North Sea – without, apparently, any attempt being made to rescue them. The fact that the ship had run on to a sandbank within twenty-seven miles of Harwich, and within sight of three lightvessels, made this seeming lack of any activity all the more surprising.

The ship in question was the North German Lloyd's *Deutschland* – a 2,898-ton emigrant vessel commanded by Captain Edouard Brickenstein. She had left Bremerhaven at 9.30 pm on Sunday 4 December 1875 on passage to New York via Southampton. On board her were a crew of 99, 3 pilots, and 113 passengers. This was Captain Brickenstein's first voyage in the *Deutschland* – though he had been employed by North German Lloyd for seven years. Throughout this period, he had been in command of one or another of the company's ships.

Early on the Monday morning (the 5th), the wind began to freshen until, some time before 4 am, it reached gale force. To make conditions even more difficult, there were sudden flurries of snow that blotted out the surrounding sea. Captain Brickenstein ordered the sails to be taken down, and instructed the chief engineer to throttle back the engines to half speed. At 5 am, he stopped the ship completely. A seaman who was up in the bows taking soundings reported 17 fathoms. Brickenstein ordered 'slow ahead'.

Nobody could have accused the captain of negligence. He had

been on the bridge all night ; he had done everything possible to verify his position - even one of the pilots, who had made the voyage down the North Sea 120 times, said afterwards that he had agreed with Brickenstein's estimate. Nevertheless, this very experienced seaman was some way out in his reckoning. At the time when the sailor was calling '17 fathoms', he believed that they had covered 140 miles. In fact, the distance was about 170. The mistake could be accounted for by freaks of wind and tide plus a small compass error. The result, however, was catastrophic. A lightship, which should have informed him his ship was approaching a sandbank named the Kentish Knock off the Essex coast, was mistaken for another that was on station to safeguard ships against the North Hinder - over thirty-five miles away to the east. Whilst Brickenstein would keep to the west of the Hinder, the Kentish Knock had to be passed on the eastern side. One moment, the winking light was plain to see on the port side ; the next, a sudden blizzard blotted it out. Then there was a slight shock. Brickenstein signalled 'full astern' to the engineers, but the sudden impact sheared off the propeller. Instead of retreating, the liner was caught up by the force of the waves, and deposited farther on to the bank.

Everything happened quickly at first. Rockets were sent up, hopefully to inform the lightship's crew of the ship's plight. The snow storm made it impossible to see them ; in any event, it is doubtful whether any action would have been taken. The Trinity House crews did not necessarily pay a great deal of attention to these distress signals. They were quite frequently used by vessels to call for a pilot when heading for the Thames. But, in any case, there was a limit to the amount of information they could send ashore. Although the electric telegraph had been in existence for some time, and a cable had already been laid across the Atlantic, nothing had yet been done to link these vessels with the coastguard stations. Some ingenious character had once suggested the use of carrier pigeons ; they had been tried out, but the experiment had to be deemed a failure. The birds had been so terrified by the wind, that they simply clustered around the lamp.

Furthermore, even if it had been possible to convey the news to

Harwich, there was a limit to what could have been done. The first essential in such a situation is a lifeboat. The port had possessed one of these rescue ships – forty years previously. It had been removed and the station closed down for lack of use.

The *Deutschland,* then, was clamouring for assistance, and nobody was heeding her. Indeed, the first news the public received of the disaster was on the following day – when a ship's boat with the *Deutschland*'s name on it was washed up at Garrison Point on the Isle of Sheppey. On board was a quartermaster named August Bock, and two corpses. One was the mortal remains of another sailor ; the other, the body of a passenger. When Bock was taken to the Duke of Clarence Hotel in Sheerness, he had a grim story to tell.

Before the liner hit the sandbank, apparently, the lead had been cast every half-hour. The water was becoming more and more shallow, but still the *Deutschland* crept on towards her doom. Once she went aground, Brickenstein had given orders to prepare the boats. One was launched with the fourth officer in charge and eight people on board. A large wave swamped it almost immediately, and all the occupants were drowned.

Bock was given charge of the second boat to go over the side. On the first attempt, it turned over. It was, however, attached to a three-inch rope, which prevented it from being swept away. Somehow, they managed to right it, and Bock and his two companions jumped in. With such a strong sea running, there was no time to wait for anyone else. In any case, the rope suddenly snapped, and they were swept away. They managed to hoist a small sail ; and, as they surged past the lightship, they called for help. Nobody seemed to have heard them.

The passenger, who was already injured, died at three o'clock that afternoon. By nightfall, the other seaman was obviously suffering from the effects of exposure. Bock rubbed his hands in an endeavour to keep his circulation going, and tried to revive his spirits by telling him that they were only a short distance from land. Sometime during the night, however, the man died. Had Bock any idea of what might have happened to the remainder of the people aboard the ship ? He had not – though he was not hopeful. He suspected that he might be the only survivor. When he completed his account, he remarked 'I hope to have a better night's sleep than I have had these past three nights.'

Quartermaster Bock's boat was, in fact, the only one to get away from the *Deutschland.* The onslaught of the sea was increasing ; Captain Brickenstein was wondering whether it was safe to make another attempt, when a series of waves decided the matter for him. As they broke over the ship with fearful violence, the remaining lifeboats, which could have accommodated all the passengers, were, one after another, torn away and slung into the ocean. The only hope, now, was assistance from the shore ; and, despite the distress rockets that seemed endlessly to soar into the sky ; despite, too, the firing of the ship's gun, no help came.

At first, there seemed to be no great danger ; for the ship appeared to be standing up to her ordeal. The real peril was the rising tide. If only assistance reached them before it was too late, they might all be taken off safely. Soon after striking the bank, the *Deutschland* had slewed round, so that the waves were now breaking over her side. Captain Brickenstein's immediate fear was that they might be carried off the sand and into deep water. The hull had clearly been damaged by the impact ; if this happened, she would surely sink. To prevent it happening, he put down an anchor.

She was now making water in the forward compartments. Until it ran out of steam at 6.30 that evening, the donkey boiler was able to drive the steam pumps. But the passengers had been on the verge of panic ever since they realized what had happened. It would be better, Captain Brickenstein decided, if the men were given some task to take their minds off their worries. He put them to work on the hand pumps. Their morale improved, though the women were still in a very nervous condition.

The hours dragged by. The gale showed no signs of moderating, and there were some more falls of snow. Two steamers were seen ; more rockets soared into the sky, but without any effect. The vessels kept on course and showed no signs of being aware of the stricken *Deutschland.*

Whatever the *Deutschland* may have lacked, it was not supplies of food and drink. Meals were served frequently – again, to act as a distraction. The passengers moved around below decks in their life-jackets, dimly aware of the desperate race that was taking place outside. On the one hand, there was the rising tide ; on the other, the hope of rescue from the shore. With every minute that passed, the likelihood of a victory for the tide increased.

Round about nightfall, there was a moment of hope. More rockets had been sent up, and these were answered by the Sunk lightship away to the south-east. The Sunk passed a message on to the Cork, which flashed a signal to the coastguard at Harwich. Now, surely, there should be some action.

But there was none. Sixty-year-old Captain Carrington of the tug *Liverpool* was, as it happened, in the coastguard station when the message was received. The *Liverpool* had steam up and could have set off for the wreck immediately. But Carrington said that it was impossible. With a high wind, visibility cut by the incessant snow storms, and night approaching, there was nothing he could do. Any rescue attempt would have to wait until morning. His mate, a more venturesome mariner by the name of King, disagreed with him. There was a heated argument, but the captain's views prevailed. He had, after all, spent forty-six years of his life at sea. He should know. The *Liverpool* remained moored to the quayside.

At Ramsgate, where a lifeboat was stationed, the distress rockets had been seen. A policeman passed the news from the coastguard station to the harbour master, and the crew was alerted. But they took the same view as the *Liverpool*'s master. Under these conditions, and with the wind coming from the north-east, it would be impossible to reach the area of the wreck – let alone find it and to take off the stranded victims.

On board the *Deutschland,* the situation was deteriorating. The deck was now awash ; and, at two o'clock in the morning, Brickenstein ordered everybody up from below. Some meekly obeyed ; others preferred to remain in their bunks and die in comparative comfort. The next hour or so was dreadful. As the water level climbed higher, so these refugees had to climb on to the roofs of deckhouses or into the rigging. The shrieks and sobbing of the women were described as 'agonizing'. When a seaman made his way along the main deck in an attempt to rescue a child, he was caught by a wave and smashed against the bulwark. He was killed instantly. Captain Brickenstein had a narrow escape when the bridge was carried away. Luckily, he managed to clutch on to the rigging and save himself. Five nuns, who were on their way to America from a Franciscan convent in Westphalia, were less fortunate. The senior member of this small flock was a woman about six feet tall. For some time, she had been crying 'Oh Christ – come quickly !' Then, suddenly, she was silent. She and her companions had been washed overboard.

Now the effects of the cold were beginning to strip refugees away from their positions in the rigging. The purser was one of those whose hands became too numb to grasp the shrouds. He fell into the sea. Brickenstein kept on reassuring people that they would be safe, if only they could hold on. The fate of the purser had shown how difficult this was. Two men decided to greet death on their own terms. One hanged himself behind the wheelhouse ; the other slashed his wrists. And the water was still rising. Now it was scouring its victims off the tops of the deckhouses. The rigging was the only hope ; but it was a very slender one. Of the two hundred people who sought sanctuary there, fifty were washed away. Three men and one woman, who had refused to go on deck, were drowned in a corner of the engine room.

During the night, the lights of two ships were seen passing by. One was a 1,000-ton steamer on passage to London ; the other was a brig that remained in sight for four hours. Neither seemed to be aware of the tragedy that was taking place only a short distance away.

At some time, round about dawn, the wind swung round to the west-south-west and moderated. There were no more snowfalls; and when Captain Carrington looked out across a slate grey sea at 8.30 that morning, he could see no more reason to keep his ship tucked up snugly in harbour. With a crew of sixteen on board, he gave the order to cast off. One-and-a-half hours later, he reached the wreck. By this time, the tide had gone down again, and all those who were still alive had been allowed down on deck. They were cold, dazed, and still frightened – but nobody seemed to be seriously ill. The arrival of the *Liverpool* brought hope at last; but a heavy swell was running, and one and a half hours went by before the cautious Captain Carrington judged conditions to be safe enough to transfer survivors. The tug's boat was used for the operation; on the first trip, it brought off nine people – on the second, eight. The last to leave were Captain Brickenstein, his first, second and third officers, the ship's doctor, and three of the engineers.

Afterwards, the recriminations, led by *The Times,* began. Nobody could understand how a large liner could be wrecked within such comparatively close distance of the shore – and thirty-six hours elapsing without an attempt at rescue. Carrington was described as a man whose age was 'nearer 70 than 60' (untrue) and accused of cowardice. His lack of action, which might have been the result of senility, was one of several facts that, said *The Times,* had 'excited universal indignation'. And why, the newspaper also wanted to know, had the Ramsgate lifeboat not put to sea? Over the previous sixteen years, its crew had made 240 trips and saved between 400 and 500 lives. Why had they been inactive on this occasion?

But there was another, and greater, mystery. Apart from the passing ships and the light vessels, there had been other boats in the vicinity. These were smacks, some of them quite sizeable craft, owned by the fishermen of Harwich and Ramsgate. To judge by the reports that were circulated afterwards, it seemed that their crews had acted in the manner of vultures – waiting until the sea had battered all life out of the *Deutschland*, and then going aboard it to pillage. The fact that many of these men owed a considerable portion of their incomes to salvage work, gave substance to the rumours.

An artist employed by the *Illustrated London News* visited the wreck a day or so after the disaster. When his drawings were published, one of them showed a dastardly scene in which the smacksmen were mutilating corpses in the saloon and stripping it bare. According to the caption, the villains were Harwich fishermen. One man was cutting a hole in the deck ; another rifling a gladstone bag. A depraved-looking character with a drunken leer on his face was drinking from a bottle and waving a lady's fan. Nearby, a companion was sorting through a suitcase of clothes. And so on – it was a situation fit to shock the most broad-minded of the magazine's readers.

The Mayor of Harwich promptly wrote to *The Times*, protesting that no Harwich men had been anywhere near the wreck. On the following day, the editor of the *Illustrated London News* replied that the Mayor's son had accompanied his artist on the trip and had witnessed the plundering. Finally, the Mayor gave vent to his conviction that 'the men of Harwich are not cowards', and that the whole unfortunate episode merely showed that the town should have a lifeboat.

Captain Carrington was vindicated by several experts, who agreed that his course of action (inaction, perhaps ?) had been the right one. Nothing could have been done that night, and he would simply have been jeopardizing the safety of his tug for no good purpose. The directors of North German Lloyd seem to have agreed with this view ; at all events, they sent Captain Carrington a letter of appreciation. As for the Ramsgate lifeboat, everybody was satisfied (even *The Times*, eventually) that it would have been impossible to make the trip of thirty miles or more in the face of a north-easterly gale – even if a tug had been available to tow it.

Eventually, the recriminations petered out. The survivors of the *Deutschland* were taken to Southampton and sent on their way to America in another ship. Their lives had been saved ; but, so far as these hapless emigrants were concerned, that was all they had left. All their possessions had gone down in the *Deutschland* ; they now faced the prospect of establishing themselves in another country without so much as a spare suit of clothes.

4 COLLISION COURSE

Rescue is a transitive verb ; it requires an object to make sense. When the cross-channel steamer *Seaford* came into collision with a freighter named the *Lyons* on a summer's day in 1895, it was a very private affair. One ship was dispatched to the bottom of the sea, but not a single life was lost. Nor, indeed, was anyone seriously hurt. But this immaculate conclusion to a disaster took place without any assistance from outside. It could not even be seen by the prying eye of another ship ; for the whole drama, from beginning to end, was played out in dense fog. Nothing could be more private than that, and nothing more orderly than the conduct of *Seaford's* passengers. The episode must surely qualify for the word 'rescue' ; but the people aboard saved themselves.

The S S *Seaford* had come into service during the previous year. She had twin screws, an average speed of 20 knots, and she was proudly referred to as the 'flagship' of her owners' (the London, Brighton and South Coast Railway) fleet. Her master was an intrepid mariner named Richard Sharp. Captain Sharp's previous command had been a paddle-steamer named the *Paris*. On one occasion, after the *Paris* had been injured in a gale, he nursed his ship back to port after a terrifying voyage lasting thirty-six hours. Such a feat gave a man a reputation, and Captain Sharp was held in very high regard - both by his employers and by passengers using the Newhaven-Dieppe service to the Continent.

At 1.30 on the afternoon of 20 August 1895, the *Seaford* slipped out of Dieppe harbour on her way across the Channel. The weather was beautiful ; the sea as smooth as a sheet of plate glass, and a brilliant sun beating down from a cloudless sky. The 255 passengers (among them a hundred Americans) were all on deck, watching the ship cleave her way towards Newhaven - some of them waving to Captain Sharp as he stood on the wing of the bridge. It was impossible to believe that they were only about two hours away from serious trouble.

Half an hour after the *Seaford* had sailed out of Dieppe, the French cargo vessel *Lyons* departed from Newhaven for the French port. She was one of those drudges of the cross-channel trade about which little is known. Nor is there much information about her master, except that he was a veteran of the Newhaven-Dieppe route, named Captain Ganvin. He too, had remarked on the fine weather and was preparing to enjoy what promised to be an uncommonly easy passage.

But the English Channel is treacherous. The weather can change suddenly, and many of its silver linings conceal clouds. The *Seaford* was two hours out of Dieppe, when she suddenly ran into a bank of fog. The *Lyons* had travelled rather less distance, when she entered the far side of the hazard. Both captains ordered reductions in speed ; both ships uttered throaty stay-away-from-me warnings on their whistles. The fog thickened until visibility was estimated at half a ship's length. The two steamers groped blindly on.

The trouble about fog is that it not only frustrates the eye ; it also confuses the ear. Sound can certainly be heard, but it is difficult to say where it is coming from, and equally hard to estimate how far away the source is. But neither captain was unduly worried ; on the evidences of their journeys so far, there seemed to be little traffic about.

At some point, Captain Sharp heard the dismal cry of the *Lyons* ; and, at about the same time, Captain Ganvin's ears picked up the sound of the *Seaford*. Both men now knew that there was an intruder on the scene ; neither could tell exactly where it was. And then, suddenly, like a pair of phantoms, the two vessels sighted each other. Captain Sharp instantly ordered a change of course, and rang down 'full astern' on the engine room telegraph. It was too late. Like a dagger approaching and then penetrating a body, the *Lyons* slid towards the *Seaford*. The effect of the latter's change of course was merely that, instead of colliding head-on, the *Lyons* pierced her side. Within seconds, water began to pour into her hull just behind the engine room.

The bows of the *Lyons* no longer had their elegant, wedge-like, shape. They were, indeed, very badly battered, and her forward compartment was beginning to fill with water. Nevertheless it seemed probable that, unless the watertight bulkhead gave way, she would be able to remain afloat. The *Seaford*, on the other hand, was manifestly destined for the bottom of the sea. And unless the signs were very misleading, this unhappy event would not be long in taking place.

As Captain Sharp quickly decided, there was no time for the usual formalities of a shipwreck – such as sending SOS signals and manning the boats. If the people on board his ship were to be saved, they would have to cross over on to the *Lyons* ; they would have to act quickly and with perfect discipline. With the aid of a megaphone, he communicated his decision to Captain Ganvin, who promised to do all that he could to help. Ganvin coaxed his vessel away from the passenger ship and then, with magnificent seamanship, brought her gently alongside.

The passengers had been given life-jackets and ordered to parade on deck. Like a well-drilled company of soldiers, the 255 passengers, the seamen, the stewards, the engineers, the stokers, and, finally, the *Seaford*'s officers stepped across the small chasm that divided the one ship from the other.

Only two mishaps occurred. When the two ships collided, a male passenger rather foolishly jumped over the side in a moment of panic. He was now clinging to the anchor chain of the *Lyons*. He had to be hauled on board. His pride, no doubt, was injured ; but it was a small price to pay for an act that could have had much more serious consequences.

A certain Mrs Pearslow, a lady who came from Southend, was the victim of the other misadventure. As she was crossing the gulf between the two vessels, she slipped and fell into the great divide. That might have been the end of Mrs Pearslow, if it had not been for the *Seaford*'s second officer. This gentleman, a Mr Moor, had already earned one medal for life-saving ; now he was to give further evidence of his courage. He was lowered on the end of a line to where the luckless Mrs Pearslow was thrashing about in the water. Gently, or so it seemed, he took hold of her, and the two of them were hauled to safety.

Captain Sharp preferred to linger for a few last moments on board his ship. He collected some documents from his cabin ; then, when the *Seaford* gave an ominous lurch, he judged it wisest to depart. Walking with great dignity across the deck, he climbed over the rail , and jumped into one of the boats that were waiting below. Shortly afterwards, he joined Captain Ganvin on his bridge.

There are times when the sea can be downright theatrical. This was one of them. No sooner had Captain Sharp gained the bridge of the *Lyons*, than, quite suddenly, the fog peeled away from the scene and the sun came out. The *Seaford* slowly raised up her bows as if in salutation ; paused for a second or so ; gave (or such seemed to be the case) a quick sigh ; and then slid gently and uncomplainingly into 50 fathoms of water.

Lyons was a midget compared to the more majestic *Seaford*. There was little room on board for all this sudden influx of passengers and sailors, and many of them had to be accommodated in the hold. So long as the bulkhead held, they were safe. If the weight of the water on the far side stove it in, the *Seaford* would quickly have a companion on the sea bed.

The cargo steamer turned back for Newhaven. The bulkhead withstood the water ; and, at eight o'clock that evening, she deposited her load of humanity on the quayside. The passengers had lost all their travelling belongings in the *Seaford* ; and this, in many instances, meant their money. There was, however, an American senator among them who had been more prudent. This man, whose name was Ozmum (it should be recorded, for he deserves to go down to posterity), had an uncommonly large amount of cash in his pocket book. As soon as he stepped ashore, and having discovered the plight of his travelling companions, he called for a table and chair. Then, in the soft twilight of that summer's day, he opened an *ex tempore* banking service. Can self-help ever improve on this ?

At about the same time, some of the other passengers tried to open a subscription for Captain Sharp and his colleagues. They had, as every one agreed, performed their duties exceedingly well. Not surprisingly, the amount collected was rather small – though, as the captain realized, this was not quite the point. The point was that they were thinking kindly about him, and this must have been comforting. Captain Sharp needed a little comfort : he had just lost £50,000's worth of ship.

5 RADIO TO THE RESCUE

The 1,964-ton freighter *R F Matthews* was an undistinguished ship ; a slattern of the seas that prowled from port to port in search of cargoes. No artist painted her portrait, no writer included her in *Great Vessels of Our Time*, or anything like that. She was a nonentity. Nor was the accident that occurred to her on the evening of 28 April 1899 remarkable. She was outward-bound from London in thick fog, when she rammed the East Goodwin lightship ten miles offshore from Deal. No lives were lost ; neither vessel was badly damaged. The episode had all the makings of a non-story.

Had the *R F Matthews* collided with any one of the three other lightships that marked the limits of the Goodwin Sands, the matter would have soon been forgotten. Had she, for that matter, rammed any other vessel, the *R F Matthews* would have continued her journey into obscurity. But, nudged by chance into the East Goodwin, she earned herself a place in posterity. Of all the guardians of shipping off the UK coast, this particular example was unique. She was equipped with wireless – that amazing invention by Guglielmo Marconi, which was about to revolutionize communications. What is more, this was to be the first occasion on which a vessel in distress had ever sought assistance with the aid of Marconi's discovery.

The first person to receive news of the accident was a gentleman named Bullock. Mr Bullock, who was Marconi's assistant, had established an experimental radio station on shore at the South Foreland lighthouse. He could, if he wished, hold conversations in Morse Code with the lightship. Earlier that year, a merchant ship named the *Elbe* had run aground on the Goodwins.

The master of the lightship, Captain Clayson, had seen her distress signals, and he tapped out a message to Mr Bullock - who then summoned the Ramsgate lifeboat. In the present instance, however, the lightvessel herself was involved ; and, according to the first message, the situation was serious.

On the evening of 28 April, then, Bullock was inside the lighthouse - tinkering, perhaps, with his equipment. Suddenly, the alarm bell sounded. He put on his earphones, and picked up the East Goodwin's signal. The message read : 'We have just been run into by the steamer *R F Matthews.* Steamship is standing by us. Our bows very badly damaged.'

Mr Bullock immediately dispatched a message to the Trinity House authorities at Ramsgate, and a rescue operation was begun. The lifeboats at Ramsgate, Deal and Kingsdown were alerted. The Trinity House vessel *Warden*, which was in the vicinity of Beachy Head, was instructed to proceed immediately to the lightship's assistance ; and, since the damage sounded serious, arrangements were made to replace the injured vessel.

Meanwhile, Mr Bullock was busy tapping away at his transmission key. Would Captain Clayson give him some details of the occurrence ? Had, for example, there been any casualties ? No, it appeared that none of the seven hands aboard the East Goodwin had been hurt - though they were badly shaken. The vessel had been struck broadside on, and had very nearly been sliced in two. This was to say, explained Captain Clayson, that her stem had been smashed right down to the water line. A strong tide was running, and visibility was down to a few yards. Yes, the *R F Matthews* was still standing by.

In fact, the damage turned out to be considerably less than the messages had suggested. The *R F Matthews* resumed her voyage, and the East Goodwin remained on station. But, as a portent of radio's role as a life-saver, it was impressive. Afterwards, Captain Clayson was full of praise for what, initally, he may have regarded as a new-fangled gadget. 'I had not the slightest trouble,' he said. 'I called the South Foreland up about two minutes after the steamer got clear of us, while she was still standing by, and I got the reply immediately from Mr Bullock, to whom I telegraphed briefly the particulars of the collision.'

An unprepossessing ship, an event of no great importance ; on some occasions, just such trivialities make up history.

The better part of twelve years were to go by between the episode near the Goodwin Sands, and the next time that wireless was used for rescue at sea. On this occasion, in 1909, the matter was of far greater importance. Indeed, without the aid of Marconi's invention, a good many lives might have been lost. Once again, the cause of the catastrophe was fog. The White Star Line's 15,377-ton liner *Republic* was outward-bound from New York for the Mediterranean. On board her were 250 first-class passengers, 211 in the steerage accommodation, and a crew of 300. The value of her cargo was estimated at half a million pounds ; it included 500 tons of supplies for the U S Mediterranean fleet. They were urgently needed. Sicily and Calabria had recently been devastated by earthquakes ; the warships had been sent to assist the refugees, and they had parted with large quantities of food and clothing. The stores in the *Republic*'s holds were intended to replace them.

On 22 January 1909, a thick fog settled over New York and the sea and coastline round about. It was so dense that ferry boats travelling between Manhattan and Staten Island had difficulty in navigating. A number of ocean-going captains had decided to delay their departures until the weather cleared ; their vessels were lying helplessly at anchor in Upper New York Bay.

Shortly after midnight, Captain Scalby, master of the *Republic*, decided that he could wait no longer. Groping his way out to sea, serenaded by the deep notes of his ship's whistle, he passed the lightvessel at the approach to the Ambrose Channel, and headed into the Atlantic. Coming towards him, nearing the end of a passage from Naples, was the Italian liner *Florida*. Most of the eight hundred emigrants on board her were victims of the earthquake disaster. Having lost their homes and most of their belongings, they were planning to make a fresh start in America. At 5.30 on the morning of 23 January, the two vessels collided.

When, some days later, the matter was investigated by a court of inquiry, the *Florida*'s captain asserted that his own ship had been proceeding at reduced speed, whilst the *Republic*'s master had made no concessions to the weather and was travelling flat out. Captain Scalby, not surprisingly, denied this ; but, whatever the truth, it was too late to take any evasive action when the two liners came within sight of each other. The *Florida*'s bows crashed into the *Republic*, and it soon became evident that the White Star ship was in serious trouble.

Captain Scalby's first reaction to the collision was the traditional one of all mariners in distress – to send up rockets. Before long, however, he realized that this was a waste of time. With visibility reduced to a matter of yards by the fog, nobody would be able to see them. Then he awoke to the fact that, on board his ship, he had a priceless piece of apparatus designed for just such an emergency – wireless. The impact of the crash had put the dynamos out of action, and the wall of wireless officer Jack Binn's cabin had been shattered. But the accumulators were undamaged, and Mr Binns was able to transmit. The message began with the letters CQD – a signal devised by Marconi, and intended to mean 'All Stations – Urgent'. In fact, most people were misled by the letters and interpreted it as 'Come Quick – Danger'. For this reason, it had been replaced by SOS in the previous year. Perhaps Mr Binns was a creature of habit : all accounts of the disaster report that he transmitted CQD.

But, at the time, nobody was inclined to argue about the niceties of wireless procedure. The important thing was the message, which read 'Am in distress and need assistance.' It was quickly answered by the French liner *Lorraine* with the encouraging words, 'I am coming.' Since she was about 120 miles from the scene, her captain, Commandant Tournier, passed the word on to a Cunarder named the *Lucania*. The *Lucania* was, he believed, nearer. Commandant Tournier also advised Captain Scalby to 'Make all the noise you can.' At first, this was not quite so helpful as it might have been ; the Nantucket lightship was somewhere in the vicinity. She was sounding her fog horn and ringing her bell. At least one would-be rescue ship mistook the sounds for those of the *Republic* and went to the wrong location.

On board the badly damaged White Star liner, everything was proceeding commendably calmly. Indeed, according to one witness, if anybody was looking for panic, he would do better to study the Italian ship, which was standing by in case the *Republic* sank. All the *Republic*'s passengers had been asleep at the time. Some of them had been flung from their berths by the impact. As soon as they realized what had happened, they rushed on deck in their night clothes. Four people had, in fact, been killed - and four more had died in the *Florida*. But the situation did not seem to be particularly serious. The well-disciplined crew were reassuring ; and Captain Scalby made it clear that there was nothing to worry about. Soon, they would be transferred to another ship - probably to the *Florida* which, beyond having the shape of her bows changed, did not seem to be very badly damaged.

The early morning air was now alive with the chatter of Morse Code. Two stations on shore received the signals, and relayed them. Jack Binns was working in complete darkness ; one wall of his radio shack was now missing, and his transmitting key was damaged. But, for so long as the aerial remained in position and there was life in the accumulators, he would maintain his supply of information. One message was sent to the radio station at Vineyard Haven (at the entrance to Nantucket Sound) for transmission to the White Star office in New York. It read: 'When 20 miles south of Nantucket lightship this morning, we were run into by an unknown vessel. The engine room is full of water. Can remain afloat. No danger. No lives lost.'

Its inaccuracy was not confined to that phrase 'No lives lost'. In fact, the *Republic* would not be able to remain afloat for very much longer. On the credit side, however, was the fact that several eager helpers were hurrying to the liner's aid.

The steamer *New York* came out to escort the *Florida* into New York. The White Star liner *Baltic* was inward-bound from Europe. At 6 am, just as she was passing the Nantucket lightship, her master, Captain Ransome, received news of his colleague's plight from the wireless station on shore. He promptly turned his ship round and went to the *Republic*'s assistance. The *Lucania* was still searching; and two US coastguard cutters, the *Gresham* and the *Seneca*, were racing to the disaster.

Appropriately, the *Baltic* reached the *Republic* first, and the passengers were transferred to her in a well-disciplined exercise that was remarkable only for its lack of incidents. At one o'clock that afternoon, they were landed in New York - in, it seemed, surprisingly good order. As one observer noted, 'It was hard to tell the difference between those who had been rescued and those who had not.' However, he commented on 'the worn and somewhat unkempt appearance of the ladies'. And, from another, 'Barring general mourning for the loss of baggage, everybody is in good spirits. It was wonderful to have the wireless at hand, otherwise it would have been a terrible calamity.' Later, the good spirits were to vanish, and the mourning for the baggage to become more pronounced. Having escaped with their lives, the indignant survivors crowded the White Star offices, uttering angry threats of litigation for the loss of their practically all.

Out at sea, the drama was approaching its conclusion. Captain Scalby, wireless officer Binns, and a volunteer crew of officers and ratings were still on board the *Republic*. The two coastguard cutters had attempted to take the maimed liner in tow - and, for a short period, it looked as if they might be successful. At eight o'clock that evening, however, the *Republic* seemed about to settle. The cutters launched boats to take off the men who had stayed behind. Scalby and the chief officer were picked up clinging to a grating, as their ship slid to her death in 45 fathoms of water off Martha's Vineyard.

Captain Scalby and his wireless officer arrived in New York to heroes' welcomes. When they stepped ashore, they found a gathering of White Star stewards and seamen waiting for them. These men hoisted them on to their shoulders, and, cheering and blowing trumpets, carried them to the White Star offices. The two officers seemed to be embarrassed, but they deserved the praise that was now lavished upon them. Nobody could have put it better than an editorial in one newspaper, which paid tribute to 'The men of steady nerves and courageous hearts who, patiently and skilfully, directed the splendid rescue - and who are willingly sharing the thankfulness and gratitude of the country with Mr Marconi, whose wireless telegraphy made it possible.'

Before the year was out, Mr Marconi's wireless telegraphy was to come to the aid of another liner in distress. The ship in question was the 10,606-ton Cunarder, *Slavonia* ; and the sorry truth is that - had it not been for the disobedience of her commander, Captain Arthur George Dunning - the *Slavonia* need never have come to grief. The unfortunate affair cost Captain Dunning his master's certificate, so let us not be too hard on him. Nevertheless, the truth is that he deliberately ignored his owners' instructions. One of those mysterious and hitherto unsuspected currents (which were so often used as explanations as to why a certain vessel was not where she should have een, and were a captain's last resort at a court of inquiry) was present - or said to be. It may very well have nudged the *Slavonia* playfully in the wrong direction. But the fact is that this fine ship had no business to be anywhere near where she was on 11 June 1909.

In making the passage from North America to the Mediterranean, the shipping companies of England and Germany took entirely different attitudes. The Germans favoured a course that routed their vessels via the Azores for the very sensible reason that it saved time. The English owners - notably Cunard - considered that the Islands represented an unnecessary hazard and should be given a wide berth. If only Captain Dunning had been more obedient, he would have retained his command, and the *Slavonia* would have lived to make many more crossings of the great ocean.

The trouble was, perhaps, that the captain had too many ideas. The voyage from New York to the *Slavonia*'s intended destination at Naples was long and tedious. Having spent so many days looking out at the grey and heaving monotony of the Atlantic, the passengers would, Dunning decided, welcome a change of view. What nicer surprise could they have had than to wake up one morning, and find themselves regarding the green and pleasant shapes of islands serene amid the sunlit sea ? Far from adding to the length of the trip, it would, if German experience were anything to go by, make it shorter. Economies in time and distance equal money saved. On the one hand, Captain Dunning would receive the grateful thanks of his customers ; on the other, he would be rewarded by his thankful owners.

Had it not been for the treacherous current, the presence of which was confirmed at the court of inquiry by the master of a Harrison liner, Captain Dunning's notion might have worked. He had intended to pass within three miles south of Flores, the westernmost island in the Azores group. Later, when conditions became hazy, he modified this to nine miles. As it happened, he might have done better to stick to his original plan ; for, when the moment of reckoning came, he was ten-and-a-half miles north of his estimated position and heading straight for land. As it happened, nobody would have found much to enjoy in the view no matter what had happened, for Flores was wrapped in mist. Indeed, even if the weather had been perfectly clear, the green and pleasant lumps of land would have passed unseen ; for, when the *Slavonia* found her progress blocked by an all-too-solid reef of rocks, the time was getting on for midnight. Captain Dunning's collection of errors was not confined to his ship's position ; his sense of timing appears to have been faulty as well.

The *Slavonia* had sailed from New York on 3 June. Most of her first-class passengers were wealthy Americans from the mid-west bound for holidays in the Mediterranean (what with the misfortunes of the *Republic* and the *Slavonia*, it seems to have been a bad year for United States citizens intent upon visiting this part of the world). Her cargo was made up of coffee, machinery, lubricating oil, typewriters, tobacco and three motor-cars.

For the next seven days, the *Slavonia* steamed on across the ocean. For most of the time, the skies had been clear ; Captain Dunning had been able to take good observations and, thereby, to establish his position. Round about 10.30 on the night of the 9th, they encountered patches of fog, but they were nothing to be alarmed about. There was a look-out in the crow's nest and another in the bows. Between them, these men should have been able to spot any approaching trouble in plenty of time. Captain Dunning maintained the liner's full speed of 13 knots. Next day, the mist had cleared away.

It was on 11 June, that the *Slavonia* touched land for the last time. Since it was unexpected and unseen, she touched it rather too violently. The moment she was stopped short with a considerable jolt, Dunning gave the orders for full speed astern. But then he realized that he had not connected with a gentle bank of sand, but with a sharp and remorseless chunk of rock. The liner was already taking in water up front, and she might well be past the point of no return. The safest thing would be to beach her. The engine room telegraph bells jangled again – this time, ordering full ahead. She lurched forward and brought her bow up on to what was to become her final resting place.

The first action was to send out a distress signal on the ship's wireless. It was nothing if not succinct : '*Slavonia* ashore, please assist.' This was followed by particulars of her position, and that was all. No details were given. Once it had been sent, Captain Dunning ordered ten of the ship's boats to be lowered. The night was calm ; there seemed to be no urgency about anything now that the damage had been done. People who came on deck were assured by Captain Dunning that the vessel was unsinkable. Something similar was to be said three years later to travellers in the *Titanic*. In 1909, people still believed such things ; in this instance, of course, it was true. With her fore part firmly resting on the edge of an island, little short of the supernatural could have thrust the *Slavonia* under the sea.

In latter-day airlines, a common way of taking the minds of nervous passengers off the perils of flight, is to give them plenty to eat and drink. So long as the jaws are moving, it appears, fear is forgotten. Captain Dunning anticipated this theory in 1909, by ordering the cooks to maintain a continuous supply of meals. When it became light, the passengers were offered trips ashore in the lifeboats. Later, he ordered a gang of seamen to rig a rope from the front of the vessel to the top of the cliffs that overlooked it. There was no particular purpose behind the exercise, he afterwards admitted, but it gave the passengers something to watch. It may also have diverted their attentions from the after-part of the vessel, where, with the rising tide, the stern was becoming submerged – and, amidships, the water was dangerously close to the level of the bridge.

The *Slavonia*'s distress signals had been picked up by the Eastern Telegraph Company's station on the Azores, and by two ships. The first to receive them was the German liner *Prinzess Irene*, which had left New York two days after the Cunarder's departure; the other vessel was an American liner, the *Batavia*. Both changed course and headed for Flores. The *Prinzess Irene* arrived first. Although she had been 180 miles away when she received the signal, she covered the distance remarkably quickly, and soon her comforting shape could be seen anchored off the island.

Within a few hours of the *Slavonia*'s going aground, the inhabitants of Flores had begun to take an interest in the wreck. When they saw that the passengers would have to be ferried out to the German liner, they offered the use of their boats. There was, however, a limit to the number that the *Prinzess Irene* could take. She, too, was bound for Naples, and her accommodation was booked to capacity. Since, it was reasoned, a first-class traveller pays more for his passage, he deserves to be rescued first. Consequently, the steerage passengers were asked to await the arrival of the *Batavia*. Their better-off companions were transferred without incident to the German ship, which then departed for Gibraltar, where they were landed.

Presently, the *Batavia* arrived, and the steerage customers were put on board her. In case there was any feeling of unfairness, or in case this seemed to suggest that the *Batavia* was a second-rate ship, it was quickly pointed out how fortunate these travellers were. She had, it was explained, better steerage accommodation than the *Prinzess Irene*.

Twelve hours after the German liner had dropped anchor, the crew were the only people still on board the *Slavonia*. By now, the sea was becoming rough, and the liner's stern was sinking lower in the water. Captain Dunning made up his mind that any salvage operation was out of the question, and that the apparently impossible might be about to happen. If things went on like this, it would not be long before the ocean hauled his ship off the ledge of rock, and deposited her in deep water. He ordered his men ashore

– using a bosun's chair and the rope that had been rigged up to amuse the passengers. Now that the two rescue ships had departed, there could be no knowing when they would be taken off the island. But Captain Dunning was not, presumably, in any hurry. As he must have realized, nothing but trouble lay ahead.

6 ON THE ROCKS

The night of 17 March 1907 was a brute. A south-westerly gale was hurling itself up the English Channel ; the coastline of Devon and Cornwall was wrapped in fog ; and, to make matters even worse for the incautious mariner, sudden squalls of rain rendered bad visibility even worse. It was an unforgiving night, when the margin of error was nil. For any captain who mistook his position, who came close to the shore, when he should have been way out at sea, the rocks that litter the beaches were as lethal as a minefield.

Two master mariners miscalculated. The results of their mistakes were that a couple of liners were slammed on to the rocks. In terms of human life, the disasters were inexpensive. For the captains, however, the price was damnable. At subsequent courts of inquiry, they were both found guilty of negligence. In each case, the penalty was that their masters' certificates should be suspended for three months. For one of them, this was particularly humilating. It was his last voyage in an extremely competent career. He had risen to command one of the crack vessels in his owner's fleet. Now, when he should have departed from the sea in triumph, he was banished in shame. Whilst his liner was eventually salvaged and repaired, nothing could mend the captain's damaged pride.

March 17 was a Sunday. The Elder Dempster liner *Jebba* was on her way home from the West Coast of Africa. She was due to call at Plymouth ; afterwards, she would back-track around the coast of Cornwall, and proceed up the Irish Sea to Liverpool. Her commander, Captain J. J. C. Mills, was huddled in a corner of the wheelhouse, continually glancing anxiously to starboard. If only the fog would clear sufficiently, he should be able to see the loom of the Eddystone light. That would enable him to fix his position ; it would act as a signpost, pointing the way to Plymouth.

But the fog did not clear ; nor could Captain Mills or any of his crew hear the lighthouse's fog signals. It was not, perhaps, surprising. Torrents of foaming water were crashing over the liner's bows. The wind, in full chorus, drowned all other sounds except those of the giant waves and the groan of complaining girders within the vessel's hull. Nevertheless, the *Jebba* plunged on at 11 knots - which, in view of the fact that Captain Mills had only a very approximate idea of his position, was far from prudent. He was rushing when he should have been groping, unaware that he had already passed the approaches to Plymouth ; and that, if he continued on his present course, he would collide with the coast of Devonshire at a point named Bolt Tail - a headland some twelve miles to the east of the port.

At one-minute intervals, the *Jebba*'s siren bellowed a warning to other shipping. Most of the seventy-nine passengers - sixty of whom had embarked at West African ports, the remainder of whom had come aboard at the Grand Canary - were up and dressed. Some were ready to go ashore when, as they expected, the liner dropped anchor in Plymouth Sound. They were blissfully unaware that anything was wrong. But so, unfortunately, was Captain Mills.

It all happened quite suddenly. One moment, the ship - or so it seemed - was out in the Channel, cocooned in fog and battered by wind and sea. The next, she was being hurled on to the rocks at the foot of Bolt Tail. There were hastily shouted orders, the desperate jingle of engine-room telegraph bells ; but it was far too late to save the *Jebba* from what now seemed to be certain destruction. A few minutes earlier, when a seaman taking soundings had reported a depth of 11 fathoms, Captain Mills had ordered the speed to be reduced to $5\frac{1}{2}$ knots. But that, too, had been of no avail. From the moment, fifty-five minutes earlier - when he had mistaken the *Jebba*'s position - the course to disaster had been inevitable.

The passengers were herded into the saloon. Heavy seas were now sweeping across the vessel. The cabins had been flooded within three minutes of striking. Now, the floors of the public rooms were awash. One or two of the ladies fainted ; but a missionary on board was of stronger stuff. With the water up to the top of his calves, he calmly sat down at the piano and embarked on a repertoire of popular pantomime tunes. If death was on its way, at least they would all perish to the sound of cheerful music.

Up on deck, the night was pitch black. Even though the liner was perched beneath it, it was impossible to discern the gaunt shape of the headland. There was, however, no mistaking the breakers that were throwing themselves at its base. There seemed to be a very real danger that, before very long, the vessel might break up. Nor was that the only fear. Amid such dark and chaotic conditions, would anybody see the distress signals ? The crew and passengers of the *Jebba* seemed to be most terribly alone - although within yards of dry land, they were isolated in a capsule of destruction.

In fact, the signals had been observed by the coastguard, and the lifeboat from Hope Cove had put to sea within minutes of the initial disaster. But the wind was too strong ; the waves, too high. The craft could not get near enough to be of any assistance. The *Jebba*'s lifeboats were swung out ; the missionary's impromptu concert continued its melodious progress. And, on the bridge, Captain Mills and his officers studied the murk ahead for any signs of life.

After about an hour, somebody saw lights moving about on the headland. Presently, a green one and then a red one appeared. Somebody shouted, 'They're going to send a rocket.' The aim was perfect. Dragging a trail of flame in its wake, this harbinger of deliverance landed amidships. The second mate grabbed the line attached to it ; with this link to the security of land, the survival of all those on board the *Jebba* seemed less uncertain. A rope was now attached to the main mast ; and, before very long, a breeches buoy was mercifully carrying the stranded passengers and crew to safety. The ladies and the children went first ; they were set down on a ledge that gave access to Ramilles Cove - an inlet near the foot of Bolt Tail, from which they could walk safely away from the wreck. Before long, however, a second line of escape had been despatched to the *Jebba.* It was used by the men, who were hauled to the clifftop. By eight o'clock in the morning, everybody - including the ship's cat - had been taken off. What was more, the weather was moderating, and the liner was still intact.

Now they were able to land the hundred bags of mail and some of the passengers' baggage. The rest of the cargo - which consisted mainly of palm oil - would have to wait. It seemed unlikely that, having survived the initial ordeal, and with a light breeze replacing yesterday's gales, the *Jebba* need be a total loss. Nevertheless, it was obviously going to be a very difficult task to pull her off the rock. Eventually, however, a team of tugs dislodged her, and she was towed to Southampton. For everybody except Captain Mills, the story had a reasonably happy ending. At the court of inquiry, it was decided that he had 'not navigated with proper and seamanlike care'. The suspension of his certificate for three months was his penalty for having been, as he asserted, moved seven miles north-north-east of his intended course 'by an unusual current'. It was also a severe setback to his career at a time when it had seemed to be most promising ; for this was his first voyage in command of the *Jebba.*

If the disastrous voyage that ended on Bolt Tail had been Captain Mills's first (and last) experience of commanding the *Jebba,* the journey that brought the White Star liner *Suevic* so near to destruction that same night, should have marked a peaceful end to the career of Captain Thomas Jones. After a lifetime spent at sea, Captain Jones was about to retire. As a kind of finale to his umblemished record, he had been given charge of the 12,500-ton liner that was the largest and fastest on the company's service between the UK and Australia.

The *Suevic,* too, was homeward-bound. She had called at Cape Town ; and, on 13 March, at Tenerife. On board her, as she steamed into the English Channel, there were 382 passengers, 1 stowaway, and a crew of 141. Among the 1,000 tons of freight in her holds, there were the carcases of sheep, rabbits, and a consignment of butter.

Like the *Jebba,* she was battling against heavy seas, and the wayd way ahead was blanked out by fog. At about 10.30 that night, one of the look-outs reported the beam of a lighthouse on the southernmost tip of the Lizard. Captain Jones went out onto a wing of the bridge to note its position. It came from a point higher than he had expected ; they must, he decided, be too close inshore. He gave the order 'hard a' port', but it was too late. A strong current was running ; and, before the liner had a chance to answer to her helm, she was pitched forward on to rocks. One and a half minutes had elapsed since the first sighting of the light. Later, a passenger recalled :

> I was in my bunk, when I heard a grating noise. We were told that the vessel was ashore. I rushed out and could see the glimmer of the lighthouse lights ahead. The crew responded to the orders of the captain nobly. They gave us life-jackets, and we were told to go into the library. Hot coffee and other warm food was brought to us, and the liner's boats were then launched. We were let down over the side, one by one. It was terrible in the darkness, and we had to wait for the lifeboat to be hurled near the ship before we could get on board. The lifeboat crews worked bravely, but the weather was awful.

The hardest task was that of carrying young children down the ladders. Sailors took them in their arms, but it was difficult to hold on to them. Some had a narrow escape when, on its way from the ship, one of the lifeboats nearly capsized.

Once the *Suevic* had hit the rock, Captain Jones wasted no time. He immediately gave orders for the minute guns, which signalled a vessel's distress, to be fired. In a part of Cornwall notorious for its shipwrecks, there was plenty of help on hand. The lifeboats at Lizard, Cadgwith and Mullion were putting out to sea within minutes of the disaster. Presently, they were joined by two more from farther afield ; but, as in the *Jebba*'s case, the heavy seas made rescue work extremely difficult. However, a telephone call (confirmed by a telegram) from the coastguard had alerted tug owners in Falmouth. Before long, the tug *Triton* steamed through the murk. When she returned to port, she had 140 of the *Suevic*'s passengers on board.

Had there been sufficient light to see it, the rescue operation would have been an impressive sight. As the traveller observed, the discipline among the crew was faultless ; but so was that of most passengers. The women, especially, are said to have behaved well. Indeed, the only people who incurred any criticism were a clergyman and two salvationists. In spite of the inexorable rule about women and children first, they managed to jump the queue and find seats in the first boat that got away.

But they were the exception. Bravely, uncomplainingly, these intrepid travellers who, in some cases, had been snatched from sleep, and who had all been rudely wrenched from the comforts of a luxury liner, huddled in shivering clusters on deck. Then, when their turn came, they clambered down rope ladders suspended over the ship's side. The majority of them were landed at Cadgwith, where they found the village alive with hospitality.

The appalling weather made the business of bringing the passengers ashore a slow one. The next morning was well advanced before the last boatload clambered on to the beach. But, despite the time it took, everything had gone well. No lives were lost ; a few people suffered bruises and all of them were drenched, but nobody was seriously injured. A great deal of effort was now applied to the task of making their situation as comfortable as possible.

Whatever receptions the wreckers of Cornwall may have prepared for their victims in the past, the villagers around the Lizard peninsula in the year 1907 were impeccable hosts. They supplied food and clothing and fires and hot drinks by the gallon. As soon as officials of the White Star company heard about the disaster, they made arrangements to take over an hotel in Falmouth. But the *Suevic* had been on passage to London, and it was to the capital that most people wanted to go. The Great Western Railway rose manfully to the occasion. Volunteers with cars – some of them from as far away as Penzance – thronged the roads to provide what *The Times* described as a 'service of road motor-cars' to Helston station. A special train, complete with restaurant car, was waiting ; once everybody wishing to use it was on board, it set off post haste for Paddington. The adventure was over. Some of them, indeed, arrived at their destination sooner than they might have done had the *Suevic* not run aground.

Back at the Lizard, however, another rescue operation was beginning. Captain Jones and his crew had remained on their ship. The portholes and the watertight doors had been closed immediately after the impact. The pumps were busy, but the liner did not seem to be in any grievous danger. It might, surely, be possible to refloat her - even, perhaps, to tow her to a dockyard. When, on Tuesday, the sea moderated, the *Suevic*'s condition was described as 'comfortable'.

Later that day, however, the weather worsened and the liner developed a marked list to port. By now, she was surrounded by tugs, barges, and all sorts of craft. The Liverpool Salvage Association had agreed to attempt to save the liner. That afternoon, the Association's salvage vessels, *Ranger* and *Linnet,* arrived on the scene. As the sea boiled up into a maelstrom once more, it became impossible to get near the stricken ship. Indeed, towards evening, there were doubts whether she would be able to survive this latest ordeal. But they need not have worried. The *Suevic* appeared to be indestructible. Hard as the waves might batter her, she still held firm.

It soon became clear that it would be impossible to refloat her as she stood. A good deal of damage had obviously been inflicted up front. Furthermore, the job of hauling a reasonably large liner (the *Jebba* was a good deal smaller) off a not-inconsiderable rock is not an easy one. On 28 March, over a week after the calamity, a strong westerly swell signalled a return to chaos. The conditions were certainly not suitable for salvage and, sure enough, the hawser attached to one of the steam tugs snapped. The liner's condition now seemed to be critical. Water was spreading into other compartments, and the general opinion was that she might break up at any moment. It was too rough for the crew to use the boats ; but, somehow, they had to be brought to safety. They eventually achieved it by crawling, man by man, along a rope attached to the other tug. The fact that the elderly Captain Jones managed to accomplish it must have been something of a miracle.

An atmosphere of gloom came over the scene. One hundred of

the men concerned with the salvage work returned to Liverpool. The end of the unfortunate *Suevic,* which had endured so much, seemed to be only a matter of hours - or, even, minutes - away. But the following morning dawned more brightly. The liner was still resting snugly on her perch ; the sea had moderated once more and there were glimpses of sunshine. If the message of the 28th had suggested defeat, that of the 29th once more awakened hope.

There was no longer any possibility of refloating the ship intact, but there now seemed to be another solution. She was 550 feet long. If divers could manage to cut the after-part away - at a point just in front of the engine room and one-third of the way from her bows - at least two-thirds of the vessel might be saved. After that, it would be a comparatively simple matter of instructing her builders, Harland and Wolff of Belfast, to manufacture a new bow.

It was an extremely difficult job. At first, the teams of divers were unable to get inside the hull, consequently, they had to do all their work from the outside. Somehow, they had to drill a series of holes up either side of the vessel. Once this had been done, each would be plugged with dynamite. When the dotted lines of explosives were detonated, they would slice the fore-part away from the other two-thirds. That, at any rate, was the theory. Whether it would work was another matter. Quite apart from its technical feasibility, it was uncertain whether the sea, in its present capricious mood, would allow it to succeed.

However, on Monday 1 April things seemed to be going well. Reuter's man on the spot reported that, with a bit of luck, the undertaking might be completed that day. In fact, he was misinformed. An explosion did, indeed, take place, but this was not the end of it. Before the after-part could be refloated, there would have to be another. In this case, it was hoped that the turbulence created by it might assist the charges of dynamite.

Nevertheless, Monday's big bang had been impressive enough. Red flags were flown before the detonation took place, and the sirens of the tugs warned away unwary sightseers. 'Suddenly,' according to an eyewitness, 'there was a terrible explosion -

hurling debris full a mile out to sea, and sending up a great column of foam.' A piece of chain struck a crewman on one of the tugs half-a-mile away, but he was not seriously injured. One man, nearer the wreck, was harmed by the blast, and had to be taken to hospital.

In terms of, so to speak, the first slice, the operation was successful. One of its results, however, was unfortunate. The line of dynamite plugs had been set in a refrigerated compartment just forward of the engine room. The sea flooded in and released some of the dead sheep and rabbits which, presently, were washed up on the shore of the Lizard. As they began to putrify, the stench became abominable. Local labourers volunteered to remove the decomposing debris, but they demanded £2 a day. The rate was thought to be too high, and the men's offer was declined. Since this part of Cornwall had become popular as a health resort, the local authorities were bombarded by complaints. Eventually, they had to yield, and the rotting remains were removed.

At first, the huge and helpless shape of the *Suevic* was hailed as an attraction by the visitors. Excursion steamers were hurried into service, and thousands of tourists set out from such places as Falmouth and Fowey to view the wreck. But their money was wasted. It was as if the liner resented this intrusion ; as if, in her shame, she preferred to hide. At any rate, the sea and the weather guarded her against prying eyes. A heavy swell built up ; and, at the same time, a curtain of fog came down. The trippers, who had endured the discomfort of seasickness, found that, when they arrived at what should have been a remarkable spectacle, there was nothing to be seen. Green-faced and discontented, they were taken back to port.

By 3 April, the divers had managed to enter the hull. The earlier explosions had sliced through the sides ; now they were able to set a row of charges across the *Suevic*'s bottom. The previous two days had seen a return of the rough weather. Now, with calmer conditions, they were able to complete their work. First came the crash of the explosion ; the column of water that, like an enormous fountain, shot into the air. Seconds afterwards, according to an eyewitness, 'we heard the loud sounds of grinding and snapping. The after-part of the *Suevic* began to rise and fall buoyantly in the swell, showing unmistakably that the sundering of the vessel had been achieved. When the vessel parted, she floated almost exactly on an even keel. The cleavage appeared to be as if it had been accomplished by a gigantic guillotine.'

The fore-part of the liner, including the bridge, remained on the rocks ; the after-part, four hundred feet long, drifted clear. The tugs *Ranger* and *Herculaneum* took it in tow, whilst the *Blazer* and *Linnet* brought up the rear. As quickly became apparent, this part of the ship had suffered hardly any damage. It floated on an even keel ; and, when it was taken to the dry dock at Southampton for examination, the liner's own engines were able to assist the tugs. Some weeks later, the surviving section of the *Suevic* was returned to its birthplace at Belfast, where Harland and Wolff built a new fore-part.

When the pride of the White Star Line's Australian service went aground on the Lizard, there were many who predicted that this would be the end of a fine ship. In fact, she had many more years of service ahead of her. During the First World War, she served as a troopship. Later, she was sold to a Norwegian firm, who converted her into a whaling factory ship. In 1942, when it seemed inevitable that she would fall into German hands, her crew scuttled her. It was an honourable end.

But every disaster produces at least one victim. In this case, it was Captain Thomas. This very experienced mariner had spent thirty-nine years at sea. For seventeen of them, he had been in command of one ship or another ; like Captain Mills of the *Jebba,* he was a member of the Royal Naval Reserve. And then, suddenly, on his last trip, his honourable record was demolished. At the court of inquiry, the members returned a verdict that, 'The vessel was not navigated with proper and seamanlike care after 10 pm on March 17th last.' As in the case of Captain Mills, his master's certificate was suspended for three months. In practical terms, it meant nothing. Captain Jones was heading for a life of retirement. Psychologically, however, it must have been devastating. Perhaps the worst feature about the wreck of the *Suevic* was that it broke her captain's heart.

7 ORDEAL BY FIRE

On Thursday 9 October 1913, the Cunarder *Carmania* was battling her way eastwards across the Atlantic in the teeth of a gale. When she was built at the end of the nineteenth century, she and her sister, the *Caronia*, had been the pride of their owners' fleet. Since then, their supremacy had been usurped by those elegant giants, the *Lusitania* and the *Mauretania.* Nevertheless, 'the pretty sisters', as they had been called, were still above average in size, comfort and in speed. Each had cost between £600,000 and £700,000 - which, by late nineteenth-century standards, was a very considerable investment. Moreover, they were the first members of the Cunard fleet to be equipped with steam turbines. In reasonable conditions, with their engines going flat out, they could travel at 22 knots.

But these conditions were not reasonable. Over the past three days, the fury of the wind had been building itself up to storm force. By now, the liner was almost exactly half-way across the ocean. She was rolling heavily ; her hull battered by violent waves - her majestic proportions reduced to apparent frailty by the enormity of the angry sea. It was, according to her commander, Captain Barr, one of the worst crossings he had ever known.

Early that morning, the ship's wireless stuttered out a message. Seconds later, the radio operator reported to Captain Barr in the chartroom. A liner named the *Volturno* was on fire seventy-eight miles away. Barr called up the chief engineer. The number of stokers on duty was doubled and the vessel's speed was increased to 20 knots. She was now steaming straight into the wind.

Some time before the *Volturno* came into view, her position was marked by a pall of black smoke on the horizon. Presently, the stricken ship appeared. The fire seemed to have broken out at the forward end. At some point, her captain had obviously attempted to abandon ship. Five of the lifeboats were missing. The tackle from one at the after-end had, according to a message received by the *Carmania*'s radio operator, wrapped itself round the propeller. She was now unable to move under her own steam.

The more robust members of the *Carmania*'s passengers had come on deck to watch the proceedings. Among them was Arthur Spurgeon, manager of Cassell's publishing house in London. Mr Spurgeon was on his way home from a business trip to New York. 'Alas,' he wrote on the following morning, 'what we expected to be nothing more than an exciting incident developed into a great disaster, which broke down strong men and made women weep.' Mr Spurgeon was, perhaps, taking an excessively doleful view of things.

Eighteen months previously, the White Star liner *Titanic* had struck an iceberg on her way across the Atlantic. The night had been clear ; there had been a flat calm ; and the liner had taken over two hours to sink. Nevertheless, two-thirds of those on board perished. The plight of those on board the *Volturno* seemed to be considerably worse. The fire was obviously gathering strength, and it looked as if the end could not be far off. Unlike the still waters that had closed over the remains of the *Titanic* the sea was now extremely rough ; it was doubtful whether even the most sturdy ship's boat could survive for long in it. Perhaps in the final analysis, the affair of the *Volturno* may be accounted a miracle. Of all the crew and passengers aboard, only one-third perished. The loss of life was in inverse proportion to that of the *Titanic.*

But, right now, it seemed doubtful whether anybody would escape. Captain Barr's first action was to lower a boat. The first officer, a powerful man named Gardner – who had once survived for twenty-three days in an open boat – was put in charge of it. No sooner had the craft touched the water, than the sea set about its destruction. Within a few minutes, five out of the eight oars had either been smashed or wrenched away. It now seemed probable that it would be picked up by a wave and flung against the *Carmania*'s hull. Captain Barr called down to Gardner, 'Get the crew out and let the boat go.' Three seamen got out and scaled the rope ladders up the liner's side. But Mr Gardner was still in full control. Presently the falls were attached to it, and it was hoisted back to the davits. Mr Gardner was standing in the stern. As he stepped back on deck, the assembled passengers cheered.

As an example of seamanship, it was magnificent. It was not, however, of any comfort to those on board the *Volturno.* The episode had only given proof of what they already suspected ; a ship's lifeboat stood no chance of survival in those mountainous seas. If there was to be a rescue, it would have to take some other form.

Captain Barr's intention was to get a line on board the ship. Having failed with one method, he now took the *Carmania* herself to within a hundred yards of the Volturno's stern. The wind and the sea were too much. After several attempts to fling a rope across the gap had come to nothing, he was forced to retreat. The sight of the blazing liner's passengers huddled on the poop added to the pathos - and the frustration. Far from abating, the flames seemed to be stronger than ever. It would not be long before they were sweeping along the entire length of the ship.

'One will never forget the look of dire despair,' noted Mr Spurgeon, 'that settled on the faces of the passengers huddled together on the after-part of the ship as effort after effort to save them resulted in failure . . . We knew that only a miracle could save them.' The *Carmania*'s passengers did their best to transmit encouragement. They waved and tried to yell messages of hope. The latter were swept brusquely away by the wind, but a thin display of fluttering handkerchiefs from the blazing liner showed that the gestures had been seen.

Captain Barr had instructed his radio operator to broadcast news of the *Volturno*'s disaster. Particularly, he hoped that there might be an oil tanker somewhere in the vicinity - a ship that would be able to moderate the sea by discharging part of her cargo. As it happened, there were several vessels near enough to come to the rescue. They began to arrive at four o'clock that afternoon. By sunset, there were nine ships on the scene, and the Anglo-American Oil Company's tanker *Narragansett* was on her way with a cargo of 3,000 tons of lubricating oil.

The first to appear was the North German Lloyd liner *Grosser Kurfurst.* Then came that company's *Seydlitz* – followed by the Atlantic Transport Company's *Minneapolis,* Furness Withy's *Rappahannock,* the Russian Steam Navigation Company's *Czar,* the Leyland liner *Devonian,* the Mercantile Marine Company's *Kroonland,* and the French *La Touraine.* The *Volturno* now became like a severely wounded gladiator in the centre of an arena – the rescue ships, like steel grandstands thronged with 5,000 spectators. As C. F. Hart, a senior executive of the *Daily Mail,* who was also travelling in the *Carmania,* told his newspaper, they were 'all eagerly anxious, but unable to help owing to the mountainous seas'. 'The sight,' he wrote, 'beggars description.'

Late in the afternoon, the *Volturno*'s master, Captain Francis Inch, had sent a message saying, 'Position getting desperate. Send rafts.' Half a dozen had been thrown into the sea from the *Carmania* ; but, predictably, they went wide of the mark and drifted uselessly away a good many yards in front of the *Volturno*'s bows. By 6.30 pm, large tongues of flame were seen leaping above the smoke amidships ; at about this time, Captain Inch lowered a boat in an attempt to get a line across to the *Grosser Kurfurst.* The craft was smashed to pieces against the German liner's hull. By some mixture of skill, agility, and luck, the crew were scooped out of the ocean and hoisted to safety. But the rope was snatched away into the ocean.

As the last of the dusk faded away, the *Carmania* turned on a searchlight. For a moment, it seemed as if nothing more could be done – but at least it would enable anybody who jumped into the water to be seen. Illuminated lifebuoys were thrown into the sea, but it is doubtful whether they would have served any purpose. Like the rafts, they were snatched away and served only to litter the ocean – the merest specks on the foaming whitecaps that flung themselves at the ships.

At nine o'clock, the fire appeared to have got a grip on the midships section of the ship. There was an explosion, and Captain Inch sent his last radio message. 'For God's sake help us or we perish,' it read. A few minutes later a single rocket soared into the sky. It seemed to be the *Volturno*'s final protest ; a desperate gesture before the fire burned the life out of her, and everyone aboard her was engulfed by the flames.

There was not a man on the surrounding ships who would not have helped her - had it been possible. But earlier attempts had shown the severity of the problem. Now, there seemed nothing to do but watch and hope. The best that could happen would be that the beleagured travellers would jump and, presently, be picked up. But how many of them would survive ? The spectacle of the passengers huddled on the poop had shown that there were a good many women and children among them. How would they manage in this turbulent sea into which it was not possible even to launch a lifeboat ? Their chances had to be reckoned as nothing.

Some time just before midnight, there appeared to be a slender thread of hope. Was the fire making less progress than it had done so far ? Was it possible to discern a slight moderation in the wind ? In either instance, the difference was small, but so is the margin between absolute disaster and survival. At all events, the master of the *Grosser Kurfurst* decided to take a chance. He lowered a boat with twice the normal crew on board. After two hours of struggling against the waves, it managed to get within hailing distance of the *Volturno*. But enough was enough ; it was as if the storm was determined that nothing should get close enough to be of any assistance. The boat was torn away and it was only with a great deal of difficulty that it managed to return to its ship.

Round about this time, the *Minneapolis* also launched a boat. An officer and six volunteers were on board. Just as the *Grosser Kurfurst*'s lifeboat had managed to get within earshot of the floating furnace, so did this. The officer called to the people on board to jump. He would, he promised, pick them up. If they had followed his advice, the result would have been disastrous, for there was nothing like enough room for them in this comparatively small craft. But nobody moved. They were either too terrified - or else, perhaps, they could not hear. At any rate, they remained where they were and, within a minute or so, the opportunity was gone.

The *Minneapolis*'s boat was now in trouble. A wave had broken its rudder into small pieces and, as soon became very evident, it was unable to regain its ship. Unless somebody took some very quick action, the young officer and the six volunteers would be swept away into the infinity of the Atlantic Ocean - beyond recall and beyond rescue.

Captain Barr in the *Carmania* rang down 'slow ahead' ; picked up the boat in his searchlight and closed on it. Five hours after they had departed from the *Minneapolis,* the crew was in reach of safety. The six seamen climbed up the rope ladders. The officer was the last to leave. A rope was lowered, he secured himself to the end and was promptly swept into the sea and knocked unconscious. As he was hoisted up to the deck, a wave seemed to take its cue from events. It suddenly picked up the craft and flung it against the *Carmania*'s side. In a second, a tolerably strong ship's lifeboat had been converted into matchwood.

This manoeuvre caused Captain Barr to change his plans. His boats were ready and his men were standing by - desperately waiting for any opportunity that might present itself. Now he was on the outside of the circle of mercy ships. To make matters worse, what the *Carmania* possessed in terms of speed and comfort was offset by poor handling characteristics. Barr judged that, from now on, the other vessels would be better able to deal with the situation. He signalled to the captains that he hoped they'd do the best they could. Beyond using his searchlights to illuminate the water, there was little purpose he could serve. From now on, he told them, 'I'll stand aside.'

Few of the *Carmania*'s passengers slept well that night. Some remained on deck ; others dozed fitfully in the first-class drawing room ; some gathered together and held *ex tempore* prayer meetings ; some wandered out on to the deck, watched the scene in which the brilliance of the burning *Volturno*'s flames competed against the glare of the searchlights ; and then returned to pace up and down the dining saloon. All of them were waiting for news from the ship's infirmary. Before very long, they believed, they would be hearing what, exactly, had taken place in the *Volturno.*

At 9.30 pm, the sky had suddenly been lit up as some of the *Volturno*'s woodwork caught fire. Then came the message from

Captain Inch followed by the lone rocket. Shortly afterwards, a lookout on the *Carmania* had heard a voice crying 'Help ; Help !' from nearby. The searchlight was directed at the sound ; and there, only a yard or two away, was a man struggling in the water. Two of the Cunarder's sailors were lowered into the sea on the ends of ropes. Despite their almost superhuman efforts, it took twenty minutes to get him safely on board. He was obviously in bad shape, and he was immediately put to bed. When he was able to speak, he was questioned about what had occurred. The man's command of English was not good ; but, with assistance from an interpreter, the story was coaxed out of him. It was a colourful, though somewhat inaccurate, story.

The man's name was Walter Trintepohl - a German who, for the last few years, had been employed by a firm of merchants at Barcelona. Recently, he had been offered a better job in New York, and he had bought a third-class ticket on the *Volturno.*

According to his account, the fire alarms had been sounded at six o'clock on the morning of the 9th. The passengers had been assembled on deck and issued with life-jackets. Captain Inch had been reassuring. It was not a serious blaze ; it would soon be put out ; but it was as well to take no chances. The women had been upset by this unpleasant turn of events. Many of the children had begun to cry. There were, he said, many babes-in-arms on board.

Had Herr Trintepohl any idea of what had caused the fire ? He was uncertain. He believed it had broken out in number one hold ; he thought it had been caused by an explosion brought about by the propeller shaft over-heating. Since the hold in question was in the forward part of the ship, and the shaft was in the after-part, this was a contradiction. Thereafter, it might have been wise to treat this by-now-talkative survivor's testimony with caution.

Contrary to the captain's announcement, it had been impossible to quench the flames ; and, at ten o'clock that morning, the boats had been prepared for launching. Apparently, the captain and ship's officers on board the *Volturno* were all English, and Herr Trintepohl spoke well of their conduct. His admiration for their men, however, was a great deal less. 'I am sorry to say,' he told his small audience, 'that the crew, who were Germans and

Belgians, behaved very badly. People rushed about wildly, and the crew seemed to think they ought to have first place ; and, instead of quietening, they made the panic worse.'

This view was the first report from a survivor to appear in the newspapers. Consequently, it coloured the public's attitude to the behaviour of the *Volturno*'s seamen. Not surprisingly, the German and Belgian authorities were worried by these aspersions. The Belgians ordered an inquiry ; the German reply was more devious. It was reported in the newspapers some days later, and purported to come from a German news agency. Herr Trintepohl had, according to this announcement, got it all wrong. The crew had behaved impeccably ; it was the officers who had lost their heads. 'They ran hither and thither and behaved very badly', the story read. 'They apparently did not know what to do.'

But the survivor's story was not yet done - and there were worse things to follow. The first officer had charge of the first boat to be lowered. The rule in such matters was 'women and children first'. In fact, the women and children had not been able to get near it. When it was sent down to the sea, it was filled with members of the crew. However, there was justice of a kind waiting for them. Just as it touched the water, it was hurled against the ship's side and broken into two. Everybody on board was drowned.

Was it simply bad luck - a freak of ill fortune that would not occur again ? The chief steward, according to Trintepohl, seemed to think so. At all events, he threw some provisions into a second boat and clambered on board. Once again, the men scrambled on board - leaving no room for the women and children. But this, too, was a descent into disaster ; like its predecessor, it was smashed against the side of the heavily rolling liner, and everybody was killed.

Thereafter, said Herr Trintepohl, 'all was confusion. Ropes broke and occupants were thrown into the sea and drowned and killed.' He said that six boats had been launched in all. Was it true, his questioners wanted to know, that two had succeeded in getting away and were now missing ? No, he replied. They had all met with the same fate. Indeed, the attempt to escape by this means had been so disastrous, that Captain Inch had cut the tackles of all the other boats – to make sure that nobody would try to lower them.

As soon as the *Carmania* arrived, he said, the stokers rushed up on deck from the boiler room. The engines may have been out of action, but the captain still needed steam to drive the pumps and the generators. At first, he told the firemen to return to duty. When they refused, he was compelled to produce his revolver and send them below at pistol point.

Had, then, the passengers' behaviour been any better ? Trintepohl shrugged his shoulders. The women, he said, had wept, shrieked and become hysterical. 'Some women and some men tore their hair,' he observed, 'but others were quite still.' Some, it seemed, had also jumped overboard – just as some of the crew had done. None of them had ever been seen again.

The account contained very little for anyone's comfort. To the perils of flame and storm must now be added hazards of the victims' own making. Even with a well-disciplined crew and a courageous collection of passengers, the chances would have been small enough. With panic evidently taking hold of those on board, any likelihood of a happy outcome seemed to be minimal.

However, when the night melted into another slate-grey dawn, there were things that gave the observers grounds for hope. The *Volturno,* though still blazing, was, none the less, still afloat. The tanker *Narragansett* had arrived with (as they used to call it) 'the milk'. And the storm had definitely moderated. The miracle that Mr Spurgeon had deemed necessary for survival might take place after all.

The *Volturno* had been built for the Canadian Northern Railway Company. One of the smaller members of the North Atlantic's passenger steamer population, she had cost £80,000. In 1913, she was under charter to a line named the Uranium Company. Most of her voyages were between Rotterdam, Halifax and New York. She carried more lifeboats than the regulations required ; she was considered to be reliable and thoroughly seaworthy.

On 2 October 1913, she set out from Rotterdam for Halifax.

On board her were a crew of 93 and 564 passengers. Many of the latter were emigrant families from Eastern Europe. Two hundred and sixty of them were heading for Alberta and Saskatchewan, where they proposed to build new lives in a politically and economically more settled climate.

Six days after the liner's departure from Holland, a gale sprang up. By the following day, the weather had worsened and a strong beam sea was causing the ship to roll abominably. It was this motion that, some time early in the morning of Thursday, the 9th, caused a barrel of barium oxide in number one hold to overturn. The container was broken, and the action of the chemical on the wooden deck probably started the fire. It was, apparently, first discovered by a male steerage passenger who was on his way to the women's quarters. The man yelled out a warning, and then fell victim to panic. He rushed up on deck and threw himself overboard. However, his words had been reported to the chief officer who, at 6.50 am, informed Captain Inch. Five minutes later, the flames had burned through one of the hatches, and the deck fittings on the forecastle were alight. Captain Inch ordered the chief engineer to reduce speed. He directed the quartermaster to keep course before the wind – to give the fire extinguishers the best possible opportunity.

But the fire was working quickly and with deadly efficiency. The watch below were trapped in their quarters ; within minutes of the blaze beginning, they and a number of steerage passengers were burned to death. All told, the casualties amounted to well over sixty. Soon after this, tongues of flame were reaching into the sky on a level with the light atop the fore-mast. To add to the havoc, a chain of explosions damaged the ship's compass, and the blast wrecked the saloon and hospital amidships.

Captain Inch decided that there was no possibility of saving his ship. He ordered the boats to be provisioned and swung out. The passengers were assembled ; stewards issued them with life-jackets. The ship was now rolling so heavily that the shelter deck became awash each time she swung over.

The first boat to be lowered was number two. The liner's chief officer, an old Atlantic hand named Miller, was in charge ; on board were a number of cabin-class passengers attended by stewards.* As the lifeboat touched the water, it capsized. Miller and several members of the crew succeeded in righting it. They climbed on board and were immediately swept away from the ship. No trace of them was ever seen again. The passengers were drowned.

Number three and fifteen boats were the next to be lowered. Neither reached the water; each was smashed against the hull on its way down, and its human contents spilt into the sea. There were no survivors. Number six boat managed to make it; but, like number two, it was carried off by a wave and never seen again. Finally, the action moved to the after-part of the ship, where an attempt was made to float number seven boat. It was this that fouled the propellors and put the engines out of action. But, by this time, Captain Inch had heard that the *Carmania* was on her way. He decided that any more attempts to launch boats were doomed to disaster – and, in any case, unnecessary. The exercise was abandoned. Later, he did, indeed, order the falls of the surviving boats to be cut away. But, as he explained afterwards, the object was to join the lengths of rope into a line that might be dispatched to one of the rescue ships. It was not, he emphasized, a measure intended to *prevent* the use of the lifeboats.

Three hoses were at work on fighting the blaze. At one time, about mid-morning, it seemed as if they might be successful. But then the fire, which had been quiet for a while, erupted again and burned its way into the ship's coal bunkers. By 9.30 that night, the saloon, the chart house, the bridge – and indeed, everything forward of the funnel was on fire. It was then that Captain Inch was compelled to leave his post and come aft.

Marconi operators W. Seddon and C. J. Pennington had been transmitting signals continuously since the beginning of the fire. When the engine room had to be shut down after the flames had reached the bunkers, the generators were among the casualties. Nevertheless, Messsrs Seddon and Pennington remained on the

*Herr Trintepohl's account was wrong – he had obviously not witnessed the launching of the boat. No doubt, he first saw it after it had overturned. The slanderous report on the crew's behaviour was based entirely on surmise.

air – using accumulators for electricity. It was not until an explosion carried away the aerial at 11 pm that they had to close down. Then they joined the unhappy band of travellers on the poop.

Contrary to the report of Herr Trintepohl, the crew seem to have behaved very well. At first, there was panic among the passengers. Early on, they had made an attempt to get into the lifeboats, and the sailors had been forced to drive them away with sticks. After that, they calmed down. Even so, Captain Inch clearly did not place much reliance in them; for when, during the night, he discovered that the fire was burning its way aft below decks, he took pains to keep the knowledge from them. As for the stokers, he had no complaints about their conduct. When he was questioned about them at the subsequent court of inquiry, he said, 'I never thought of taking my revolver out of my room. I never had a revolver on me.'

The night was pitch black, with sudden squalls shrieking out of the murk, hammering the ship in their fury, and then hurrying on their way. Despite the searchlights, it was too dark to make any effective attempt at linking up with the rescue ships; in any case, the weather was still too bad. Earlier on, Inch had encouraged his men and the passengers to jump for it in the hope of being picked up. Once he realized that the fire was spreading, it had seemed to be the only alternative to death by burning. The passengers, numbed by panic and terribly afraid of the ocean beneath them, refused to move. To give them a lead, the sailors were told to jump in groups of two and three at a time. At nightfall, there had been fifty-eight crewmen on board. By morning, there were only eight. Of the fifty who tried to swim to safety, thirty-two were rescued. Of the passengers, Walter Trintepohl seems to have been alone. The others clung doggedly to their precarious perch on the poop. As Inch said, 'They all seemed ready for their doom, whatever it was.'

Throughout the night, people passed the time in different ways. The captain, the two wireless operators, and the steadily decreasing crew of sailors busied themselves in constructing small rafts. Some of the women sang hymns; most of the male passengers smoked heavily. A few actually managed to snatch a few hours sleep, whilst a group of Jewish travellers held a prayer meeting lasting for several hours. The sound of their chanting could just be heard about the frenzy of the wind.

At last the long night was over. The gale had moderated, and the tanker *Narragansett* was offloading lubricating oil on to the sea. At 5.30 a.m., the Leyland liner *Devonian* managed to get a boat across; and, before long, an armada of lifeboats was in the close vicinity of the *Volturno.*

On the whole, the work of transferring the passengers and the remainder of the crew was an orderly undertaking. A rope was strung across the deck; the men were herded to one side of it, the women and children to the other. The rule was that the former were not to move until the latter space was empty. Two men tried to jump the queue. One of them threw a child into a boat and then climbed over the rope. In spite of attempts to stop him, he swarmed down a line and into a boat. Its crew, it seemed, disapproved of his action. According to one witness, 'they gave him a hammering'. Another man managed to get into the women-and-children quarter by wearing an officer's great coat. Captain Inch himself knocked the trespasser down; he was made to wait until last – and did not get away until 8 o'clock, when he was personally escorted by Inch and the two wireless operators.

One woman threw her child into a waiting boat. Her aim was bad, and the infant fell into the space between the lifeboat and the ship's hull. Under these conditions, there was a considerable danger that the youngster might be crushed to death. However, one of the *Devonian*'s crew dived in and hauled the boy out. His mother, who had been hanging on to Inch's arm throughout these moments of terrifying suspense, suddenly gave the master mariner a kiss. He seemed to enjoy it.

There had been 657 on board the *Volturno.* Of this number, 521 were rescued. The last to leave the ship was Captain Inch. After he had made a thorough inspection to see that nobody had been left behind, he made his departure at 8 o'clock – in company with the two Marconi men, the male passenger who had refused to wait his turn, and the ship's papers. The small party was taken on board the *Kroonland.* Some days later, they landed in New York.

It would be nice to report that all the troubles of those who had sailed in the *Volturno* were now over. This, unfortunately, was not the case. Ten vessels had converged on the blazing liner; the 521 who were rescued were divided among them. The score worked out as follows: *Grosser Kurfurst* - 105, *Czar* - 102, *Kroonland* - 90, *Devonian* - 59, *Seydlitz* - 46, *La Touraine* - 40, *Minneapolis* - 30, *Narragansett* - 29, *Rappahannock* - 19, and *Carmania* - 1 (the indestructible Herr Trintepohl).

Owing to the very nature of things, families had become split up, and the situation was still further confused by the fact that the liners were heading for different destinations. For example, the *Czar* was on passage to Rotterdam, the *Grosser Kurfurst* to New York, the *Devonian* to Liverpool, the *Minneapolis* to London, and *La Touraine* to Havre. The job of unscrambling the situation and reuniting the various mothers, fathers and children took ages. Indeed, one of the most pathetic figures of the catastrophe was a four-year-old little girl who turned up at Liverpool. She could speak only one word of English, and that was 'Daddy'. Over and over again, this sad-eyed mite murmured 'Daddy, Daddy!' as she wandered helplessly through the crowds. But nobody came to claim her.

The *Volturno* refused to die. Five days after the fire had broken out, her hull was sighted by Captain Arthur Rostron in the Cunarder *Carpathia.** Captain Rostron reported that it was still smoking and drifting at a speed of one knot. But it seemed to be in good order and was 'liable,' he said, 'to drift for ever.'

* Eighteen months earlier, Captain Rostron had been in command of the *Carpathia* when the ship went to the rescue of the *Titanic*'s survivors.

8 SUBSMASH

A submariner had to be (and doubtless still is) a very special kind of person. He had to be capable of living at unpleasantly close quarters with seventy or so other men for long periods. He had to be calm, resourceful, the type of man who never loses his head. An admiral who was present at the launching of one submarine, described these communities who lived and fought beneath the sea as 'exceptionally civilized'. He might have added that they were exceptionally courageous. Of all the branches of the Royal Navy, their calling was, perhaps, the most hazardous.

During the twelve years following the First World War, there were four submarine disasters. In 1924, the *L4* was rammed by the battleship HMS *Resolution* off Portland. *Resolution* survived to take part in the Second World War, and to die peacefully at the hands of shipbreakers in 1948. *L4* was damaged beyond repair. Forty-three men died in her mortally wounded hull.

In 1925, the *M1* – a boat easily distinguishable by the large 12-inch gun that was mounted in front of her conning tower – collided with another vessel off Start Point. Sixty-eight men lost their lives in her. Four years later, on 9 July 1929, the *H47* and the *L12* were on their way home from manoeuvres in the Irish Sea. At some point off the coast of Pembrokeshire, they came into collision. Twenty-four officers and ratings were killed.

It is, perhaps, significant that all these disasters occurred as the results of collisions. By the very nature of her work, a submarine is designed to live within the sea and not on top of it. When she is on the surface, she presents a low profile; she is inconspicuous and easy to overlook. That, too, was part of the idea. The present-day nuclear submarine can spend long periods submerged with no need to come above water. The earlier boats, on the other hand, were driven by electric motors. They depended on batteries for their power. From time to time, they surfaced to recharge them by means of their diesel engines (which, incidentally, also provided the motive power when they were above the waves). It was on these occasions when they were most vulnerable. Consequently, they were not designed to attract attention. In wartime, it was their strength; in peacetime, their weakness.

The P-class submarine *Poseidon* had been built by Vickers in 1929. She was 260 feet long; her armament consisted of one 4-inch gun and eight 21-inch torpedo tubes. Six of them were located in her bows; two of them in her stern. On the surface, her speed was 17½ knots. Underwater, it was 9 knots. In March 1930, she was commissioned at Barrow. Shortly afterwards, she was posted to the China station. She arrived at Hong Kong in February 1931.

At 12.45 pm on the early afternoon of 9 June 1931, *Poseidon* was on passage about twenty-one miles north of Weihai - travelling on the surface from the Yellow Sea through the straits that separate Korea from the Shantung Peninsula. There was a thick mist at the time, but her commanding officer had other things to think about. He was engaged in an exercise that involved a mock attack on another warship. He was preparing to go in for the 'kill' when, suddenly, out of the fog steamed a Chinese freighter named the *Yuta.* Neither vessel, it seems, saw the other until it was too late. The *Yuta* struck the *Poseidon* at right angles, cleaving a considerable hole in the submarine's hull. Within two minutes, *Poseidon* had plunged 120 feet to the bed of the Yellow Sea.

By some miracle, five officers and thirty ratings managed to get out of the conning tower hatch before she went down. Eighteen men, however, were still trapped inside the hull. Even with the assistance of Davis submarine escape apparatus, their chances of survival had to be reckoned as small.

The code word for a submarine disaster is 'Subsmash'. Understandably, it produces quick reactions. Within a commendably short space of time, an armada of warships was converging on the scene of the collision. The cruisers *Berwick* and *Cumber-*

land and the aircraft carrier HMS *Hermes* were among the first arrivals. The destroyer *Stormclad* hastened to the area; the sloop *Marzion* and the depot ship *Medway*. The United States Navy placed its salvage ship USS *Pidgeon* at the British commander-in-chief's disposal (Vice-Admiral Sir William Kelly). Later, Sir William himself turned up in his flagship, HMS *Suffolk*.

Unfortunately, this gathering of naval might was of little comfort to the men trapped in the dying *Poseidon*. Their problem was simply to escape from their prison on the ocean bed and to reach the surface. No assembly of warships could accomplish this. Despite the number of vessels that now marked the spot, these men were very definitely on their own. There would, obviously, be an attempt to salvage the *Poseidon*, but these things take time. Long before it had been accomplished – if it were *ever* achieved – their supply of oxygen would have run out.

The men who escaped had been picked up by the Chinese ship. As they stood on her deck, gazing at the calm, mist-wrapped waters, it is easy to imagine their feelings. It was as if they tried to will those eighteen trapped men to survive; as if, by wishing hard enough, the bland surface of the sea would suddenly erupt, and the eighteen would triumphantly appear.

But the day's stock of miracles had been exhausted. Two hours and fifteen minutes after the accident, a couple of men surfaced. Some time later, four more appeared. That was all. Nobody else emerged. Nor was this frugal offering a cause for much gratitude. Of the first two, one was unconscious when he came up; the other appeared to be in a state of extreme exhaustion. Both died shortly after they had reached the sanctuary of a warship.

The hero of the *Poseidon* disaster was a petty officer named Willis. Afterwards, he was awarded the Albert Medal in gold for his valour. He was also offered promotion to the rank of warrant officer. But, no: Petty Officer Willis did not think the rank would become him. To be made a chief petty officer would be perfectly adequate. That was all he asked, and that was what he received.

Inside the submarine, the men were trapped in two compartments. Eight of them (including two who had been killed by the collision) were in the front of the boat; ten of them in the after-compartments. It was from the former that the chances of escape seemed most likely.

Once the submarine had been struck, she gave a sudden lurch to starboard and began to sink by the bows. The leads supplying electric light had all been cut, and the only means of illumination was by torches. It had been possible to close the watertight doors, but the situation was made worse when one of the bulkheads buckled beneath the pressure of water.

Petty Officer Willis, who was a torpedo-gunner's mate, took charge of things in the forward compartment. His first action was to advise the men to say their prayers. He then ordered them to put on their Davis escape apparatus; and, careful man that he was, he checked that they all knew how to use it. Apparently they did, with one exception. This was a Chinese boy who worked as a steward. Willis gave him what must have been the quickest course in survival that anyone has undergone. At the end of it, the young Chinaman said that he felt able to save himself.

Before they could break out, however, the pressure inside the submarine had to reach a par with that outside. To do this, it was necessary to flood the compartment. It took two hours and ten minutes before Willis judged it to be safe. By this time, the water was up to the men's knees. It had become very cold, and there were signs that the supply of oxygen was running low. There was, clearly, a time limit on this business of survival.

Two men got away before the pressure slammed down the hatch. Another, seemingly endless, period of waiting followed. Then, at last, conditions were ready for a second attempt. Willis was the last to go. He took with him the Chinese steward, Ah Hai. With the exception of the two men who had been killed, the forward compartment was now empty.

But there were other men - still, hopefully, alive - in another part of the boat. Would they be able to reach safety? *The Times* was to say of the affair that, 'Never, perhaps, has there been a more wonderful and dramatic escape from death by drowning.' It was true - up to a point. But was it sufficient to merit such extravagant language? There were still those ratings at the after-end, who - unless something happened quickly - would die in a singularly horrible manner.

When the news of the disaster reached Shanghai, all the ships in the harbour respectfully lowered their flags to half mast. The master of a Japanese salvage tug that was scurrying to the *Poseidon*'s final resting place, confidently predicted that he would raise the wreck in twenty-four hours. For the moment, however, everything was in the hands of the USS *Pidgeon,* and her crew were having a difficult time of it. When divers went down, they observed a large hole in the submarine's side. They had tapped on the hull in an attempt to see whether there were signs of life in the after-compartment. There had been no reply. The men, they feared, were dead.

The weather was now deteriorating, and the divers were finding it difficult to work in a depth of 20 fathoms. On 10 June, a signal from HMS *Hermes* suggested that conditions might make it impossible to continue the work. At one time, they lost contact with the *Poseidon* completely. But the crew of USS *Pidgeon* were determined to carry on. On the 12th, the divers went back again, and they managed to get a cable round the hull. This should have been the breakthrough. On the following day, they hoped to bring up her to the surface.

But then the weather went completely to pieces. Everything had to be postponed until the 15th, when the skies cleared and the wind dropped and the divers were able to go down again. Unfortunately a good deal had been happening on the sea bed during the days between. *Poseidon*'s hull was now covered by several feet of mud. Nothing more could be done. The attempt was abandoned, and the ships – which had gathered like mourners around a grave – returned to habour. They had tried hard; but, in the end, the uncertain weather had defeated them.

Thirty-nine survivors out of a crew of fifty-one may be accounted a reasonable score. Nevertheless, *The Times* put matters in a better perspective, when one of its leader writers pointed out that, 'In spite of the relatively small loss of life, this last tragedy of the sea has about it all the horror and suspense that are inevitably felt when one of the vessels is suddenly and involuntarily plunged to the bottom of the sea, to lie there, with living souls on board, as immovable as a stone instead of darting through the waters with the agility of a fish.'

Poseidon's toll of casualties, the newspaper noted, had brought the total of submarine fatalities to 120 since the Armistice in 1918. Nor was it the final catastrophe of the between-war years. In the following year, the M2 - which had recently (and disastrously, as it turned out) been converted to carry a seaplane - foundered off Portland. In 1939, the *Thetis* dived to disaster in Liverpool Bay on the first day of her trials. The boat was salvaged and, later, fought in World War II under the name of HMS *Thunderbolt.** But only three men came up alive on that terrible day in 1939.

War creates all sorts of technical advances. Nevertheless, the years 1939-45 did not succeed in making the submariner's life less hazardous. Nor did the peril come only from enemy warships. HMS *Truculent* was completed at Barrow on 12 September 1942. She survived the war years more or less unscathed, which makes the events of 12 January 1950 all the more ironic. Not only did they occur in peacetime ; the vessel was less than a mile from the shore, and the weather was calm.

The *Truculent* had just completed a refit in Chatham Dockyard. Throughout the day, she had been carrying out trials with her crew of six officers and fifty-two ratings on board - accompanied by eighteen dockyard workers. At seven o'clock that evening, she was on the surface making her way back to the River Medway. Her position was eight miles east of Sheerness. Three minutes later, she was mortally damaged and sliding gently towards the sea bed.

News of the disaster was first transmitted at 8.15 pm by the captain of the Dutch motor vessel *Almdijk,* which was bound for London from New Orleans. The message read, 'HM Submarine *Truculent* sunk NW Red Sands tower between X4 buoy and East Piles buoy. Have picked up five survivors. Believe submarine has been in collision with Swedish ship *Divina.* All ships please keep

*The *Thetis* may be reckoned an unlucky boat. The successful salvage operation amounted to no more than a reprieve. On 13 March 1943, she was lost as the result of enemy action.

a lookout.' Later on, the *Almdijk* signalled the North Foreland radio station, 'We are engaged in rescuing people out of the water.' There was, it seemed, still hope.

HMS *Truculent* had spent the day in the Nore Submarine Exercise area near the mouth of the Thames. By the time she was ready to put back towards Sheerness, it was dark. At 7 pm, she was travelling on the surface at 10 knots, when the lights of the Swedish coaster *Divina,* on passage from Purfleet to Ipswich with a cargo of paraffin, loomed up fine on her port bow. Something about them puzzled the *Truculent*'s commanding officer. There seemed to be two reds where there should only have been one. In fact, they conveyed the fact that the merchant ship had a potentially explosive freight on board, but this extra red light should have been removed once she had passed the Nore. Her captain had apparently forgotten. What, then, did it mean ? Was she signalling that she was out of control ? Another possible explanation seemed to be that she was a dredger. In which case, she was unlikely to be moving. The lieutenant who was the submarine's captain sent below for a manual of seamanship in an attempt to solve the mystery. He altered course twenty degrees to port, but he did not consider it necessary to reduce speed. The time was now three minutes past seven.

Had he continued on his existing course, everything would have been all right. This modification had made a collision with the small tanker inevitable. As soon as he realized his mistake, he ordered the engines to be stopped. Then he instructed 'full astern' – at the same time, calling for all the watertight doors to be closed. But it was too late. Within seconds, the *Truculent* had crashed into the *Divina*'s starboard bow. She hung there for between twenty and thirty seconds. Then she gently broke away and slid fifty-seven feet to the bottom of the estuary.

Afterwards, the master of the *Divina* recalled:

> The crash came with an impact that shook us from stem to stern, and we saw the submarine sink immediately. We saw men in the water, and my crew rushed to get them out. Ropes and lifebelts were thrown to them. We picked up eight survi-

> vors and two dead men. We gave the survivors a cup of coffee – then a lifeboat transferred them to a destroyer that had arrived.

The chief officer of the *Almdijk* which, mercifully, was in the vicinity, had this to say :

> We heard somebody calling "Ahoy, Ahoy !" from the sea. The captain stopped the ship and a lifeboat was ordered out. With a crew of six in it, it went to the area of the shouts. In five minutes we found the voices – five men clinging to five lifebelts. We put them in the boat. One of them was probably the commanding officer. At first he seemed to be uninjured, but he sat with his head between his knees as if he was in pain. I think he'd broken his arm.

The men were taken to Gravesend – where, with the exception of the commanding officer, they spent the night in hospital. The submarine's captain was transferred to a destroyer that had followed the *Almdijk* up river.

Once the Dutch cargo vessel's signal had been received, Operation Subsmash was put into action. All ships and shore establishments were alerted either by radio or landline. The destroyer HMS *Finisterre* was dispatched from Portsmouth with salvage equipment on board. The frigates *Cowdray* and *Bicester* hurried to the area – the wreck-disposal ship *Damsay,* an aircraft tender named the *Ripon* and the tug *Adherent.* In addition to all these, the Southend and Margate lifeboats had arrived, and were searching the sea round about for survivors. Finally, pilots of the Naval Air Arm were told to stand by in readiness to search the area as soon as it was daylight.

Everything possible was done to save the men on board HMS *Truculent.* The only enemies now were darkness and the tide. It was a pitch-black night with no moon to guide the rescuers. An even more formidable snag, however, was the current. The submarine had gone down on an ebb tide ; and, in this part of the estuary, a complexity of mud banks produces unusually swift currents. It was possible that men might get out of the wreck ; but, once they had reached the surface, they'd be swept out to sea. There were, heaven knows, enough ships looking for them ; but, on a night such as this, and on an ebb tide, they could easily be missed.

A number of men certainly died in this way. The rescue ship *Zest* picked up six of them ; but this was probably only a fraction of the total. Of the eighty on board *Truculent* at the time of the disaster, fifteen were rescued, ten bodies were recovered and fifty-five were never accounted for.

But some survived. At the time of the collision, a number of ratings were having supper in the stokers' mess at the rear of the boat. The shock caused four plates of food to land in one man's lap. Cups of tea were knocked off the table ; the deck tilted forward as she went down by the bows. But, said one of them, 'she went down very smoothly'. It seemed to be some kind of final tribute to the doomed submarine.

The crash had torn a large hole in the *Truculent* and the water poured into the forward compartments the moment she disengaged herself from the *Divina.* The first officer, who was in charge below, ordered everybody as far aft as they could go. The safest place seemed to be the engine room, which soon became overcrowded. At one time, something like twenty-two people were crammed into this relatively small space. But at no time was there any suggestion of panic. When the time came to use the escape apparatus, according to the first lieutenant, 'we left in a very orderly way. Although we had very little air, the boys were joking.'

Apart from the engine room, the only other safe place was the stokers' mess deck. All the men in these two compartments got out. The first lieutenant wanted to get the escape operation under way as quickly as possible. For one thing, with so many squashed into such small spaces, there was a considerable danger of the oxygen running out. For another, there was comfort in the noise of ships' propellers overhead. In spite of the fact that at 7.10 pm this officer had released a wreck buoy, he was haunted by the fear that they might go away. With an empty sea above, there would be no chance of survival.

By 8.10, the engine room had been sufficiently flooded to even up the pressures. Said one man, 'We waited for the flooding up of the engine room and then just popped out one after another without any trouble at all. We stood in a queue as though we were waiting for a bus.' The journey to the surface, according to a stoker, 'seemed an eternity'. But, presently, he came up from this underwater hell and into the clean night air. He was fortunate ; he was hauled aboard one of the boats. A telegraphist was also rescued, though he was even more lucky. There was no time for him to pick up a set of escape apparatus ; he took a deep breath and entered the escape tunnel – 'hoping for the best'. The journey upwards took something like thirty to forty seconds. But, he said, 'it felt like hours'.

According to the master of the *Divina,* none of the men he picked up were wearing life-jackets ; nor was any one of them fully dressed. What was more, the oil that had seeped to the surface seemed to have affected their eyes. Nevertheless, they were in good enough shape to be transferred to the frigate *Cowdray* and taken to Chatham.

Throughout that night and the following days, they worked on releasing the *Truculent* from her resting place beneath the sea. At an early stage in the operation, the wreck-lifting vessel *Moorfowl* came on to the scene ; and, at first light on the 13th, divers and frogmen went down. They carried out the ritual tapping on the hull, but there was no response. Meanwhile, the Southend lifeboat had covered an area extending forty miles out to sea. She came back with neither survivors nor corpses on board. It was, her coxswain said, hopeless. The Margate lifeboat had been similarly engaged. When, at last, she returned to her base, she had been searching for fifteen hours. But her crew, too, had no successes to report. And nor, when they made their sweep at dawn, had the naval aircraft.

A period of three months went by before, on 14 March, the remains of HMS *Truculent* were brought to the surface. In June of that year, they were taken across the river to Grays, where they were turned into scrap.

A boat that had come through the war unscathed had died

through confusion about the lights displayed by a harmless, 643-ton coaster. It was this small miscalculation that cost sixty-five men their lives. Or, perhaps, that figure should read seventy; for, within the larger tragedy of the *Truculent*, there was a smaller one concealed. On the night of the disaster, a Lancaster bomber took off from RAF Kinloss in Scotland with a crew of five on board. Its object was to call at Rosyth to pick up divers and then to proceed to Manston in Kent. At 11.56 pm, the aircraft heaved itself into the air, circled the airfield, and then seemed to lose height. Presently it crashed into a nearby field. All the personnel on board were killed.

9 THE SMALL SURVIVORS

One afternoon during late September 1940, a Sunderland flying boat of the RAF's Coastal Command was on patrol many miles out over the Atlantic. For several hours, the crew had seen nothing beneath them but the monotonous grey of the ocean, broken here and there by a splodge of white daubed by the foaming crest of a wave. Then, suddenly, they saw it: the merest speck on an infinity of sea - but, none the less, unmistakable. The pilot flew lower, and the tiny shape began to assume details. It was a ship's lifeboat - with, it seemed, six children among those on board. He circled it a couple of times; opened one of the flight deck windows and waved. Then he turned the flying boat eastwards. At the same time, the radio operator tapped out news of the discovery.

When the occupants of the boat caught their first glimpse of the Sunderland, they were doubtful. During the past few days, they had witnessed many similar sights. On closer inspection, however, they had all turned out to be seagulls. Was this just another deception; an unkind trick brought about by too much hope and too much imagination? When the plane lost height, they were reassured; and, when it circled the lifeboat and the captain waved, they were convinced. Now, surely, their ordeal must be nearly over.

Some while later, another Sunderland flew overhead. It dived until it seemed to skim the top of the ocean. As the roar of its engines grew louder, and it began to climb back into the sky, a package was thrown overboard. The contents were wrapped in a Mae West life-jacket. They landed nearby; when they were picked up, they turned out to be cans of baked beans, tinned salmon and peaches. There was also a note. A destroyer, it seemed, was only forty miles away. She would be on the scene within a couple of hours.

At 4.30 pm, the athletic grey shape of HMS *Anthony* appeared on the horizon. She was steaming at 35 knots; a purposeful, deadly ship - but, at this moment, she was the most wonderful, the most merciful vessel that had ever steamed across an ocean. The crew seemed mostly to be comprised of jolly uncles. They were kindly men who made jokes, and who listened with compassion when the forlorn band of survivors told their stories. The dreadful ordeal of six children, the lady in charge of them, and the other people whose lives had been spared, was over. On 26 September, they landed at Greenock. The last act in the fearful drama that was produced by a German submarine, and began in the SS *City of Benares,* was concluded.

One of the first concerns of British parents when World War II broke out was to get their children to safety. Most people believed that the German airforce would act quickly; and that, before many days had elapsed, the major British cities would be smashed into smoking ruins. But the bombers did not come. For several months, it was difficult to believe that anything had changed. There were, perhaps, more people in uniform to be seen in the streets. There were clusters of sandbags here and there, air-raid shelters, but that was all. Indeed, many families who had sent their youngsters to the safety of rural districts thought about calling them home again. The authorities warned them against it. The war was not yet over; in fact, it had scarcely begun.

In the spring and early summer of 1940, the German armies over-ran the Low Countries and France. The enemy now had air bases on the coasts of the Channel and the North Sea. Britain was only a matter of minutes away. What with this and the prospect of an invasion, the possibility of a holocaust seemed a grim likelihood. In the beginning, the countryside had appeared to be refuge enough. Now people began to wonder. On the far side of the Atlantic, the United States and Canada held out their arms in welcome. There, and not in an English village, might be the sanctuary for the nation's children.

Over in America, President Roosevelt asked Congress to set aside £31,000 to establish a bureau for the care of refugee children. Some while after the first youngsters had departed for these distant shores, Princess Elizabeth (now Queen Elizabeth II) broadcast a message to them on the BBC's *Children's Hour*. 'My sister, Margaret Rose, and I feel so much for you,' she said, 'as we know from experience what it means to be away from those we love most of all. My sister is by my side, and we are both going to say goodnight - come on Margaret! Goodnight - goodnight and good luck to you all.'

The princesses' experience of being away from those they loved most of all had been to remain in Scotland at the outbreak of war, whilst their parents returned to London. There was no talk of sending them to sanctuary on the far side of the Atlantic. To begin with, however, there was considerable class distinction in the operation. The qualification needed for this long-range evacuation was a parent who could afford to pay the passage money. This, obviously, was unfair, and the Government realized it. Presently, it produced a state-aided scheme, and an organization known as the Children's Overseas Reception Board was set up. The chairman was the Under-Secretary for the Dominions, Geoffrey Shakespeare, MP. Its objects were to select suitable children from state schools; to fit them out for the trip, to organize accommodation for them on ships, and to arrange homes for them at the other end. The destinations were to be Canada, New Zealand, Australia and South Africa.

Once the scheme was announced, the board received well over 100,000 applications. Quite clearly, it would be impossible for them all to go - there was not sufficient shipping space. Nor were parents allowed to accompany them. Each party would be looked after by a small force of volunteer escorts. Most of them were teachers. Eventually, the board shepherded 2,700 'seavacuees' (as they were called) to safety. Then, suddenly, everything came to a halt. The reason was the sinking of the *City of Benares*.

September 7 1940 saw the first German bombs dropped on London. Two days later, a party of ten escorts travelled by train to

Liverpool. They were an interesting collection of people who, between them, covered an unusually wide range of experience. There was a young Roman Catholic priest, who had taught for ten years in France, and who had only recently escaped from the German occupation. There was a parson's son, who had worked in boys' clubs; a young man who wanted to go into the church, but whose hobby was motor racing (mostly in cars he had built himself) ; a woman doctor who specialized in child psychotherapy, a nurse, an artist, a musician, and four teachers. Marjorie Day, the fifty-three-year-old headmistress of Wycombe Abbey School in Buckinghamshire, was in charge of the party. The musician was a forty-one-year-old lady named Mary Cornish. Her father was a chartered accountant ; her mother, a professional singer. She had studied at the Royal Academy of Music in Vienna.

Whatever was in store for them was obviously not going to be without its perils. During the previous four months, 351 ships — adding up to 1,300,000 tons - had been sunk in the Atlantic. The last attempt to evacuate children had come to a dismal end when, two days later, the liner *Vollendam* had been torpedoed. However, it was an adventure and not a tragedy. After two hours in the boats, they were rescued by warships and brought back to safety. But that was not all. The charges of this assorted band of escorts arrived at Fazackerly Cottage Homes in Liverpool on 10 September. That night, Nazi aircraft flew over the city and off-loaded their bombs. It was a frightening experience, but some of the youngsters were already used to such happenings.

They were aged between five and fifteen. They came from such places as London, Sunderland, Newport (Mon.) and Liverpool itself - from towns and cities, that is to say, that were vulnerable to German bombing. Some of them had lost all the other members of their families. A party of five brothers and sisters - two boys and three girls - had watched their home in the East End of London being destroyed. Another group had been discovered in an air-raid shelter without anyone to care for them. Two of them had been in the *Vollendam* when she went down. They were, one might say, already battle-hardened. But, now, hopefully, their troubles were almost over. They were due to sail for Montreal in the *City of Benares.* The date of the liner's departure was Friday 13 September. Nobody seemed to find it of any significance.

When the youngsters caught their first glimpse of the liner at Prince's landing stage, Liverpool, they found her impressive. This 11,086-ton ship had been built in 1936 by a firm on the Clyde. With accommodation for 220 passengers (it included a children's playroom), her owners, the Ellerman Line, were justly proud of her. She had been intended to bring a new standard of comfort to the company's service to India. At first, however, it had been hard to get people to travel in her. An investigation revealed that the reason was she had only one propeller. The public, it seemed, gauged a vessel's speed by the number of screws. Since *City of Benares* had only one, it followed that she must be slower. In fact, she cruised at a perfectly adequate 14 knots, but travellers to India refused to be convinced. They looked at the model of her in the Ellerman Line's London office ; noticed what they regarded as a deficiency, and booked their passages with P & O. Eventually the stern section of the model had to be masked off by a plate displaying information about the ship. It was now impossible to see how many propellers she had. The public seemed to be satisfied ; at all events, the commercial prospects of the *City of Benares* improved.

When the liner steamed down the Mersey on her way to Montreal, there were 401 people aboard her. Of these 209 were members of the crew (many of them lascars), 101 were adult passengers, and the remaining 91 were children. A block of accommodation towards the stern of the ship had been reserved for the youngsters. The girls were quartered on one side, the boys on the other. In between, there was a bathroom. Each young traveller had a small suitcase containing everything he or she had in the world. Their small supplies of pocket money were kept in sacks – with the escort of each group acting as banker.

Presently, the convoy assembled. Destroyers after the manner of sheep dogs herded the ships together. From his position on the bridge of the *City of Benares*, the commodore transmitted his instructions and the escorting warships acknowledged them on their signalling lamps. Flags fluttered in the early autumn breeze ;

a thin ribbon of white foam materialized at the stern of each vessel. The long journey westwards had begun.

At first, the weather was squally. Many of the youngsters were seasick, and the escorts were busy. Some devoted themselves to nursing the casualties ; others, to keeping the more robust members of their flock amused. The artist organized a painting competition ; there were daily parades for lifeboat drill ; but, for those who could eat it, the food was the biggest attraction. They had never tasted anything like it. As the voyage progressed and the invalids became accustomed to the ship's motion, the attendance at meals became larger ; the sound of shrill chatter louder.

The atmosphere on board was cheerful, friendly. The children were particularly fascinated by a film producer, Alice Grierson, who was on board. Miss Grierson was making a film of the voyage, and her progress around the ship could be charted by the train of eager youngsters who followed in her wake. The boat drills served as a constant reminder of the perils that attended wartime Atlantic crossings. At night, many of the children slept in their clothes ; those who preferred pyjamas had their shoes, overcoats and life-jackets assembled in neat piles at the foot of their bunks. There was, perhaps, a feeling of tension, but that was all. Nobody seemed to be nervous. When they went on deck, they could see a ship on either side. Irrationally, the sight was comforting. These merchantmen, no less vulnerable though they were, seemed to be guardians.

Mines and long-range bombers were said to be the greatest hazards. U-boats, strangely enough, were discounted. The sinking of the *Vollendam* seemed to have shown that their danger did not amount to very much. If the *City of Benares* was sunk, the most they had to fear was an hour or two in the lifeboats. As those who had already experienced it pointed out, it was quite fun. Perhaps it was as well that they had so little conception of reality.

But they did not spend much time discussing the violence of war. That was something, or so they thought, that lay behind them. They were heading for Canada, and this occupied most of their converstion. Was it really the Land of Promise ? They believed so ; a place where the sun shone incessantly and where there were no bombers, no suffering – just a boundless panorama of peace and plenty. It was a lovely thought. None of the children had experienced much in the way of comfort. They came from poor homes in slum areas – where even adequate food was often a problem. But this belonged to the past. Westward, the land was so bright they could hardly believe their good fortune.

By 17 September, the high winds seemed to have blown themselves out – though the sea was still rough. They had, according to rumours, passed out of the danger zone. They were beyond the reach of bombers ; beyond the peril of mines ; and beyond the range of U-boats. That, at any rate, was how the story went. To add substance to it, they noticed that the destroyer escort was turning back. If these small grey guardians could be dispensed with, the way ahead must surely be clear. The fact that the naval vessels were governed by the amount of fuel they could carry, that they had reached the limit of their range, does not seem to have occurred to anybody. They were, one child told another, as good as there.

After dinner that night, when their charges were tucked up in their bunks, Mary Cornish, the musician, took a stroll on deck with Sybil Gilliat-Smith, the artist. The moon was nearly full. Scattered clouds drifted across its face ; now and then, a lane of silver spread towards them across the water.On either flank, the companiable shapes of the two merchantmen plodded along in step, so to speak.

The two women were discussing music. Miss Gilliat-Smith wanted to know how 'Greensleeves' went. Mrs Cornish hummed a bar or two, and then they began to sing it. When they came to the end, they sang a few carols. Then Miss Cornish said that she was beginning to feel chilly ; she thought she'd turn in. She went below to her cabin. The time was exactly 10 o'clock.

Had they realized it, there had, for the last few minutes, been an intruder in the pack of innocent merchant ships. It was unseen, sly and lethal ; it bore no name, only the designation *U48*. With the destroyers now on their way back to Britain to refuel, the *U48* was as deadly as a fox in a hen coop. Her captain could take his time, seeking out the most succulent prizes. At 10 o'clock precisely, he took aim at the *City of Benares.*

There was an explosion somewhere amidships. The liner shuddered and all the lights went out. Miss Cornish now had to make her way aft to the children's accommodation. On the way, she met up with one of the teachers and an engineer officer. Then a sailor arrived in the alleyway ; and, at about the same time, the emergency lighting was switched on. At least they could see where they were going. They had to pick their way through a pile of debris at one point ; but, when they reached the cabins, the sight was reassuring. Only one of their charges seemed to have been seriously hurt - a girl, who had been knocked unconscious. There were no signs of any panic, though the youngsters were all a good deal frightened. What, they wanted to know, had happened ? 'It's all right,' Miss Cornish said. 'It's only a torpedo.'

Moving quickly, the children were ushered up on the boat deck. Miss Cornish was about to go back to see if anyone had been left behind, when she was stopped by an officer. 'I've already been there,' he said. 'You go to your boat station.' When she reached it, she noticed that two elements were missing from a now familiar scene. The two faithful merchant ships had vanished. They, too, had fallen victim to the *U48*'s torpedoes.

Above the babble of talk and all the other sounds that attend the sinking of a ship, the captain's voice could just be heard. 'Take care of yourselves,' he said, 'and get into the lifeboats.' The wind might have moderated, but the sea was still rough. One boat capsized as it reached the water. Children screamed ; a small avalanche of rafts and lifebuoys was hurled into the ocean. At the same time, other boats were lowered more successfully. The sea round about now seemed to be littered with the black shapes of lifeboats. Now and then, a torch on one of them winked out a signal to another.

Forty minutes after she had been hit, the *City of Benares* sank. She went down with her emergency lights still shining. According to one witness, 'It looked as if she was on fire.' Of the 406 people on board, 294 died ; only seven of the 90 children survived. Describing the U-boat captain's action as 'brutal and barbarous slaughter', *The Times* pointed out that 'the submarine commander who loosed his torpedo in darkness and a rough sea knew how small the hope of rescue would be.' The German press, predictably, asserted that the *City of Benares* was really an armed merchant cruiser.

Once the *U48* had unleashed her weapons of destruction, the convoy scattered. The majority of survivors were picked up by the escort destroyers, which hurried back as soon as they heard of the attack. But, in this hasty errand of mercy, one boat was overlooked. In it were Miss Cornish, the young Roman Catholic priest whose name was Roderick O'Sullivan, a number of other adult passengers, the liner's fourth officer, a steward, some lascars, and six small boys.

They were now four hundred miles to the south-west of Iceland. The fourth officer set a course in the direction of Ireland, but it seemed highly unlikely that they would ever get there. The only means of propelling this thirty-foot boat, which now had forty-six people on board, was by five sets of handles that worked a propeller. It was hard work ; during the days to come, as lack of food and water sapped their strength, it would be impossible. To make matters even worse, the boys had left the *City of Benares* in their pyjamas. Only one of them had a coat, and only two of them were wearing shoes.

When a group of people becomes isolated under potentially disastrous circumstances, there are always one or two who become conspicuous for their selflessness and courage. In this instance, one was Mary Cornish ; the other was the steward named Purvis. The latter was, perhaps, a born leader. He knew when to chide and when to coax the children. He kept the notion of hope continually alive, and he was a pastmaster at improvization. One of his first acts was to issue everybody with blankets. Then he erected a canvas shelter over the bows, and suggested that Miss Cornish, the boys, and the now-ailing Father O'Sullivan, should occupy this space. Two of the lads found a locker there which was promptly used as a small dormitory (curled up, two could sleep there at a time).

At first the children seemed lively and cheerful. The grey dawn brought nothing for anyone's comfort ; it was as cold as ever, the sea had become rougher, and there was nothing but ocean – a vast, terrifying expanse of heaving water. This, Miss Cornish feared, might make plain their plight as nothing else had done. With the knowledge might come terror, the final and most deadly of enemies. What did one do under the circumstances ? The answer, she decided was to sing. They warbled their way through 'Roll Out the Barrel', 'There'll Always be an England', and 'Run Rabbit Run'. Having exhausted her repertoire of what might be called 'jolly songs', she went back to the beginning and the sing-song was repeated.

Round about noon, the castaways received their first meal from the steward. It consisted of one sardine on a ship's biscuit and a dipper of water per person. The boys were thirsty, but the steward was firm. Regulations required that ships' lifeboats should carry one quart of water per person. In the case of the *City of Benares*, 1.4 quarts was the amount, but it might have to last a long time. The youngsters could plead as hard as they liked ; the steward was adamant.

During the afternoon, they talked. The topic of conversation centred round a choice of evils. What, they wondered, was best : to be bombed at home, or to be torpedoed in the Atlantic ? They might have said 'worst', but there was something magnificent in the optimism of these children ; their ability to look on the bright side of any situation was almost unbelievable.

On the following morning, the wind dropped and the fourth officer hoisted a sail. The children took it in turns to act as look-out. Something well to the front of their minds must have been the midday meal. When it arrived, it was an anti-climax. With water in short supply, they found the biscuits too dry. They asked 'Auntie' (as they now called Miss Cornish) to look after them. Perhaps later . . . They never asked for them back. By the fourth day, their thirst was so appalling, that they were hardly able to swallow.

The dread of fear had now been replaced by the peril of boredom. Fortunately, Miss Cornish had read the right books, and she told the stories well. Twice a day, much to the boys' delight, she gave her recital. The programme consisted of *The Thirty-nine*

Steps in the forenoon ; *Bulldog Drummond* after the inevitable, and seldom eaten, ship's biscuit. It may have seemed to be a somewhat limited range, but the children never tired of it. And, as Miss Cornish told them, 'you're the heroes of a real adventure story'.

Day five in the ordeal was a Sunday. As a treat, Mr Purvis revealed that he had been keeping a large can of peaches in store for the occasion. The ration was one each ; the experience was superb. At one point, it seemed as if their troubles might be at an end. During the afternoon, the boy on look-out sighted a steamer. How could they attract its attention ? During her childhood, Miss Cornish had once dreamed that she was in a similar situation. She had been rescued when she took off her petticoat and flew it as a distress signal. The memory came back to her ; what had served in fantasy might do equally well in real life. But the crew of the steamer did not appear to see Miss Cornish's petticoat. The vessel altered course and steamed away from them.* As it became dark, a storm blew up. *Bulldog Drummond* was never more urgently needed. Huddled together in their cramped wedge of space beneath the shelter, Miss Cornish told the story yet again.

There can be no rules about the behaviour of people under stress. As the days went by, the lascars became surly. On the other hand, the director of a Polish shipping company, a Mr Nagorski, showed considerable kindness to the boys. One of their great regrets was that all their pocket money had gone down with the *City of Benares.* When Nagorski heard about it, he promised to make good their financial losses.

Each day was now an ordeal. The sea had calmed down, but the supply of water had run out. Some of the children were delirious ; one of them said that his feet were so painful he was afraid of going mad. One or two found relief in sucking buttons. How far had they travelled ? Nobody had the slightest idea. Once or twice, they thought they had actually sighted land. But it was a kind of mirage ; when they looked more closely, there was nothing but

*The ship's look-out may have seen Miss Cornish's signal. There was, however, a belief current at the time that U-boat crews were using lifeboats as decoys. When a vessel stopped to investigate, she became an easy target for an attack by torpedo or gunfire.

the endless, unbroken horizon. It was just like the Sunderland flying boats that turned out to be seagulls. Even the thrills of Richard Hannay and Bulldog Drummond were losing their attraction. If being adrift in an open boat was heroism, who wanted to be a hero ?

It was on the eighth day that the Sunderland spotted them. When, at last, they went on board H M S *Anthony*, they were treated to true naval hospitality. Miss Cornish was given one of the officer's cabins. She climbed thankfully into the bunk and was thinking about going to sleep, when a steward knocked on the door. He was carrying a tray with a pot of tea on it and a plate of bread and honey. The boys were treated as the heroes they undoubtedly were. Each was given, among other things, a tin of condensed milk. In their present condition, they found it too thick and sticky to drink. Then one of them had a better idea ; why didn't they take the tins home to their mums and dads ? It was unanimously accepted.

When Miss Cornish stepped ashore at Greenock, she had a streaming cold and a reaction had set in. She disembarked at 7.30 in the evening, and was taken to what purported to be the leading hotel in neighbouring Gourock. Alas - the management did not share the views of the Royal Navy on how to look after people. When Miss Cornish checked in, the first thing she was told was, 'Dinner's off'.

On the following day, however, the happy ending department got to work. Mr Nagorski made good his promise of restoring the missing pocket money ; indeed, he did better - introducing the boys to a degree of affluence such as they had never known before. The Provost of Glasgow presented each lad with a kilt, a brooch with the city's arms on it, and a book. Then, with their gifts and clutching their cans of condensed milk, they went home.

The sinking ofthe *City of Benares* brought an end to the evacuation of children to the United States and the Dominions. Six out of the ten escorts had perished ; only one out of Mary Cornish's batch of sixteen girls survived. The trouble came twenty hours after the liner had foundered, when high seas swamped a good many of the boats. But that was the nature of the

Atlantic Ocean. It could be relied upon to finish the job that a U-boat's torpedo had begun.

10 SKIPPER ALL ALONE

On the afternoon of 21 December 1951, Hamburg was shrouded in fog. At one of the harbour jetties, the steamer *Flying Enterprise* was loading for New York. Presently, the work was completed. The hatches were closed and covered with tarpaulins. At about this time, ten passengers came aboard. The freight embraced a wide range of goods including 1,270 tons of pig iron, pickled onions and gherkins, typewriters and graphite, violins and bird cages, several Volkswagen cars, a quantity of coffee and five tons of mail for the US. Some of it had been taken on a few days earlier in Rotterdam.

Late that afternoon, the *Flying Enterprise* eased herself away from her berth, and set off down the Elbe for the North Sea. Her course to the Atlantic was by way of the English Channel. She was scheduled to arrive in New York on 3 January.

The *Flying Enterprise* was a vessel of 6,711 tons that had been built in 1944 as the *Cape Kumukaki.* Her original owners had been the United States Maritime Commission. After the war, they had sold her to the Isbrandtsen Line of New York, which renamed her. For the past three years, her master had been a thickset, powerful individual of Danish extraction named Kurt Carlsen. He was thirty-seven years old. Captain Carlsen was a veteran of the sea. He had made his first voyage at the age of thirteen, when he had sailed as a cook's assistant. Since then, he had graduated to the top of his profession, and had earned a reputation for calmness, determination and fine seamanship. During his three years as master of the *Flying Enterprise,* he had made forty-four Atlantic crossings.

Apart from his career, Captain Carlsen had two interests in life.

One of them was motorcycles; he always kept a machine on board – during his spells in port, he used to take it out on to the roads and explore the neighbouring countryside. His other delight was radio. In addition to the ship's normal equipment, he had a 1-kilowatt shortwave transmitter of his own installed in his cabin. During his leisure, he enjoyed tinkering with it.

Once the *Flying Enterprise* left the mouth of the Elbe, she steered into bad weather. It became steadily worse as she steamed down the North Sea, along the Channel and into the Atlantic. Christmas Day was not a success. Most of the passengers were ill; only one or two of them felt in any mood to face the festive fare. On Boxing Day, when the gale increased even more, Captain Carlsen began to worry about the Volkswagens in the hold. They had not been crated; there now seemed a danger that they might break loose.

The storm was far ranging. High winds gusting to 90 mph swept up the Channel and caused electricity supplies to be cut off in Kent. The liner *Queen Mary* should have docked at Southampton on Boxing Day (she was scheduled to sail to New York on 30 December with Winston Churchill on board), but she was damaged and arrived seventy-two hours late. A fisherman was drowned at Land's End while hauling his boat to safety; the 6,193-ton MV *Mactra* sent out an SOS when, 180 miles off the west coast of Ireland, her steering gear was damaged. In Mid-Atlantic, the SS *Empire Frome* was in trouble. Her deck cargo of pit props had been washed overboard – her lifeboats and derricks were all smashed. The Panamanian steamer *Buccaneer* lost a propeller; nine men survived when the 115-ton wooden schooner *Nellie Bywater* went down in the Channel; and a BOAC Stratocruiser landed in the water at Shannon after two engine covers had been torn off. At Edinburgh airport, squalls reaching 101 mph were reported; it was, the meteorological authorities announced, the worse hurricane that Britain had experienced for twenty-two years.

Captain Carlsen was concerned for the safety of his ship. He decided to stop engines and to wait until conditions moderated. If anything, they became worse. During the night of the 26th, he estimated, the wind reached a steady 60 mph – and waves of between thirty and forty feet threw the vessel about with the ease of a child bouncing a ball. On the following day, a seventy-five-foot monster smashed down on the deck. There was a sound that, at first, seemed to be an explosion. A few minutes later, a seaman arrived on the bridge. There had, he explained, been a good deal of damage forward of the superstructure. Carlsen went to investigate. The man had been right. A two-inch crack extended twelve feet down either side of the hull from the corners of number three hatch.

It might be possible to repair the damage; to plug the gap with concrete and to rig steel hawsers in an attempt to prevent it from widening. In appalling conditions, the men went to work. By evening, they had completed the job, and Carlsen was satisfied. What was more, the wind seemed to be dropping. Carlsen went to his cabin and managed to sleep.

But the fury of the gale was not yet spent. By the following morning, the weather was as bad as it had ever been. At some point, another rogue wave threw itself at the agonized ship. It smashed the starboard lifeboat, flooded the engine room, and caused the cargo to shift. The *Flying Enterprise* now had no power to drive her forward; her generator was out of action; and she was heeling over to port at an angle of thirty degrees.

Said Carlsen afterwards :

> We were doing all right until dawn the next day, when another cyclonic storm came up with some terrific seas. It was nothing but sea and water all over the place. With a broken ship, I could not steer. One tremendous sea knocked her off course . . . . It was so very touching to see the way the boys took it on the chin. I only saw one man crying, and that was either through pain or fear, because he had fallen down and hurt his arm. After twenty-three years at sea, I guess we had it coming to us. We did not want it to happen to us, we just could not help it.

This seemed to be the end. The ten passengers and the forty-one members of the crew were ordered on deck. Carlsen signalled that his vessel was in difficulties, and asked any ship in the vicinity to give details of her position. Shortly afterwards, the chief engineer managed to get the engines to start, but the list was becoming more pronounced. Carlsen told the radio operator to broadcast a 'May Day' distress signal. The Norwegian tanker *Westfal Larsen* and the US Navy's transport, *General A W Greely,* replied. Both were on their way.

By dawn on the following day, the gale had lessened. The plight of the *Flying Enterprise* was, however, a good deal worse. She was now listing at an angle of eighty degrees and both her lifeboats were in pieces. The Norwegian tanker and the transport were standing by, and Carlsen decided the time had come to abandon ship. The first attempts were unsuccessful. All told, five boats from the rescue ships were either swamped, capsized or damaged. The situation improved when the *Westfal Larsen* offloaded some of her cargo of crude oil on to the turbulent water. More boats were lowered from the transport, and these fared better.

The crew and passengers on the *Flying Enterprise* lined up along the port rail of the bridge deck. The latter amounted to five women, four men and a boy. The idea was that they should jump one at a time - each accompanied by a seaman. The first to go was fifty-one-year-old Mrs Elsa Muller. She landed in the water and was immediately picked up by a boat. After that, the operation proceeded tolerably smoothly. Admittedly, one man was drowned, but all the others were taken safely off. Considering the state of the sea, and the fact that the wind seemed to have gathered fresh strength, one casualty had to be reckoned as moderate.

Just as the last lifeboat was about to return to the *General A W Greely,* the officer in charge of it noticed an omission. Everybody aboard the *Flying Enterprise* had been accounted for with one exception, Captain Carlsen. The officer turned to the mate of the badly damaged merchantman and asked what had become of him. To his amazement, he was told that Carlsen was determined to stay behind. He had, the mate said, no intention of leaving his ship. Looking up, they noticed the captain standing on the wing of the bridge - waving goodbye. There was no point in arguing; the boat veered off and made her way back to the transport. The long ordeal of Captain Kurt Carlsen was just beginning.

With the generators out of action, it was impossible to work the large transmitter in the *Flying Enterprise*'s radio cabin. However, the 1-kilowatt shortwave set in Carlsen's room functioned by means of batteries. Its range was, admittedly, limited, but Captain Neil Olson of the *General A W Greely* had been ordered to remain in the vicinity. Any messages for the outside world would have to go from Carlsen to the transport. They would then be relayed by the *Greely*'s wireless operator.

Carlsen's first message was, 'Will, of course, stay on board until she makes port or goes down.' The possibility of the former proposition seemed to be remote. However, there was no gainsaying his determination; nor, perhaps, his logic. The *Flying Enterprise* was not yet a sinking ship. He had sent away his crew and passengers as a precaution. Her captain's duty, as he saw things, was to remain on board.

Later that day, he sent the *Greely* a more detailed account of the situation. 'The ship cracked,' he said, 'all the way across the deck and down the sides . . . We tried to make repairs. I tried to steer south to get into the traffic lanes, where we had a chance of being picked up. We were doing all right until dawn [on the 28th] when another storm came up.'

When the Ibrandtsen Line's headquarters in New York heard about Carlsen's decision, the directors tried to pursuade their stubborn captain to change his mind. He tersely refused; at the moment, it seemed, he had more immedite decisions to make. One of them was how to get from the bridge to the store room for supplies of food. With the vessel leaning over at such an angle, the journey was horribly difficult; the bulkheads now had to be treated as if they were the deck - the deck as if it were a bulkhead. For part of the trip, he managed to walk in a crouching position; at other times, he had to crawl. But, after what seemed to be hours, he made it. The storeroom was in a fearful mess; the deck (or was it the bulkhead?) was littered with food. Among the items in this jumble of groceries, he discovered a large cake that the chef had baked in anticipation of a New Year's party. This, a few cans of provisions, and some tomato juice would have to suffice. Clutching his stock of nourishment in his arms, he set off on the return journey. This time, it was even more difficult.

Since the radio set in his cabin was now his only means of communication, he decided to make the room his headquarters. He rigged up a bed in one corner where the bulkheads and the deck formed an angle. It was impossible to work the ship on his own – quite apart from the simple fact that one man cannot act as engineer, greaser, helmsman, look-out and navigator, the damage wrought by the storm had made her impossible to handle. When, a couple of days later, he was able to get round the vessel on a tour of inspection, he noted that, whilst she was now listing at sixty-five degrees, the propeller was out of the water. The only thing to do was to make himself as comfortable as he could and to await the arrival of a tug. He was told that a vessel named the *Turmoil* was due to arrive from Plymouth on New Year's Day.

Kurt Carlsen was nothing if not practical. Under the circumstances, it might have been tempting to settle for a little light reading – something that might take his mind off his dangerous and extremely uncomfortable situation. But he realized that, in the wake of any disaster at sea, there followed a squad of lawyers. When this happened, one of the points certain to be raised concerned salvage money. If *everybody* had departed from the *Flying Enterprise,* there might be quibbles about the details, but the principle would be beyond dispute. How, then, did matters stand when one person elected to remain on board? It would be as well to see how the law stood in this respect. With a candle for illumination and the chef's cake for nourishment, he curled up in his cramped corner with Martin J. Norris's *The Law of Seamen.* By all accounts, he found it absorbing reading.

The nights were dreadfully cold. After a short breather, the gale resumed its violence – and, on 1 January, the first disappointment occurred. It seemed that the tug *Turmoil* had been delayed. She had only just left Plymouth and she could not be expected for three days. However, there was some comfort from his more immediate surroundings. The *General A W Greely* had been replaced as stand-by ship by a destroyer named the *John E*

Weeks. During a lull in the storm, the warship's captain, Commander W. L. Thompson, decided that it might be possible to get a line across to the *Flying Enterprise.* Once this had been achieved, Carlsen could be sent supplies of hot food and drink. During the course of the operation, the two ships nearly collided. But, after many attempts, they succeeded. Even then, however, the difficulties were not over. Several consignments fell into the sea until, at last, Carlsen felt compelled to radio, 'I'd suggest you wait until the weather abates. I don't want the items so badly. You fellows are taking a worse beating than I am'.

But Commander Thompson, too, was a determined man. He persevered until, at long last, Carlsen had been supplied with hot coffee, fried chicken, cigarettes and a stock of candles. That night, Carlsen wrote in the log:

> What really made me decide to stick by the ship? I looked over the situation once more, and then came to the conclusion that I could manage to bring the *Enterprise* into port. I could see that she would be able to stay afloat for quite some time. I felt it was my duty.

On 2 January, he sent the destroyer a message that, he requested, should be transmitted to the owners. It stated the situation with admirable economy, 'Situation morning of the second. Number three hold full of water. Engine room taking some water, however – not much. List about sixty-five degrees . . .' If her captain was holding on to his ship, the *Flying Enterprise* was clinging no less doggedly to life. By all that was reasonable, she should have sunk.

That evening, the *Turmoil* arrived; she was a day earlier than had been expected. The vessel had been designed as a wartime rescue tug – the purpose of which was to bring crippled ships back to port. Among her successful missions had been work on a tanker that had broken in two somewhere in the Irish Sea. The rear section was taken in tow and brought to safety. The *Turmoil* was 205 feet long; her 4,000 bhp diesel engines gave her a speed of 16 knots; and she had a range of 17,000 miles. The towing line could be extended to a length of 2,000 feet. Her master was a veteran from Hampshire with fifty years at sea to his credit. His name was Daniel Parker. During the war, he had commanded a trawler employed on minesweeping duties.

Captain Parker's first problem was that of getting a line across to the *Flying Enterprise.* Somehow, Carlsen would have to manage to stand on a steeply sloping deck made slippery by spilt oil, catch the line, haul it on board, and then cope with the two-inch wire towing hawser that was attached to it. Normally, something like ten or, even, twelve men were required for the task. Would it be possible for one man to accomplish it? At night? And with the ship still pitching heavily in the rough sea? Carlsen said that he would take a chance.

He took up a position on the starboard side (the high side) of the ship; clutched the rail with one hand, and prepared to catch the line with the other. The *Turmoil* closed in - narrowing the gap until Captain Parker dared not come any nearer. Then a man up front threw the line. On the first occasion, Carlsen missed it. On the second, too, the line went wide of its mark. On the third, however, things went better. Clinging on to the rail, Carlsen grabbed grabbed at the writhing length of rope - and held it. He swiftly made it fast, but then the two ships were thrown apart by a large wave. The line broke.

There was obviously no point in continuing through the hours of darkness. It would now have to wait until dawn. With the *Turmoil* and the destroyer both in attendance, Carlsen felt that enough had been done for one day. He settled down in his corner with *The Law of Seamen.* Presently, he fell asleep.

During the night, there had been another lull in the storm. At daybreak, it returned with renewed ferocity; the sea was now corrugated into tall steep waves that caused the *Flying Enterprise* to pitch violently. Conditions could hardly have been worse for the job in front of them; but, somehow, it had to be accomplished. Throughout the morning and into the early afternoon, they made attempt after attempt. On one occasion, Carlsen caught the rope and it seemed as if the steel tow line might follow. But, as before, the vessels were suddenly jerked apart and it snapped.

By late afternoon, Carlsen was exhausted. It had been too optimistic to hope that one man, on his own, could accomplish the operation - at any rate, under these conditions. Somehow, the *Flying Enterprise*'s captain would have to be assisted. At this point, the *Turmoil*'s mate, a twenty-seven-year-old officer from Kent, came on to the scene. His name was Kenneth Dancy; he had been in the Merchant Navy since he was sixteen; and he had joined the tug a few days before she sailed from Plymouth.

The idea was breathtakingly audacious. Using all his very considerable skill as a seaman, Captain Parker proposed to steer the *Turmoil* sufficiently close to the *Flying Enterprise* for Dancy to jump from one vessel to the other. The tug approached the freighter cautiously. At one moment, the two were actually scraping together. Then they parted. During what seemed to be the fraction of a second, Dancy jumped. It was not until he was aboard the *Flying Enterprise,* that he realized he had forgotten to put on a life-jacket.

Again, the last of the light was draining out of the sky. The next attempt would have to wait until the following day. That night, Carlsen wrote in the log, 'He came on board. It was a tremendous relief for me to have a helping hand, a companion in my loneliness.' When Dancy, who had bedded down in another corner of the cabin, woke up on the following morning, Carlsen offered him a cup of tea. The water was boiled in an empty tomato-juice can held over a candle.

Before getting down to the business of linking up the two vessels, Carlsen and Dancy decided to clear away the grime and grease from the deck. They accomplished it by using, surprisingly enough, the fire extinguishers, which seemed to act as a detergent. Now, with the deck many times less slippery, they were ready to try again. Two men succeeded where one had failed. The tow line came aboard; it was secured by a metal shackle weighing 100 pounds. Before long, the *Flying Enterprise* was under tow at three knots with 750 feet of cable marrying her to the tug.

The destroyer *John E Weeks* was now replaced by the *Willard Keith.* The destination of this small band of vessels was Falmouth - 375 miles away.

Never, for an instant, was the anxiety lessened. The water sloshing about in number three hold kept on shifting the balance of the ship. It threatened to widen the crack caused by the giant wave - if it deteriorated too much, the *Flying Enterprise* would split in two. There had to be continual tours of the ship to make sure that the hawser was not breaking. One of the dangers was that it might fray. To safeguard it against this, it was necessary to lubricate it at regular intervals. Eventually, a tin of grease was sent over from the destroyer. Until it arrived, they had to make do with butter. On one of these errands, Carlsen was nearly swept overboard by a giant wave. And, as if to add to the seemingly bottomless supply of worries, a renewal of the gales was forecast.

The new outbreak of violence arrived on 8 January, when the little procession was about eighty miles from its destination. Again, the wind reached hurricane force. At first, the speed of the *Turmoil* and her charge was reduced ; then they were compelled to heave-to. For five hours they drifted on the unquiet sea - the injured ship pounded by wave upon wave - until it seemed that, at last, she must die. But the *Flying Enterprise* refused to give up ; the vessel seemed to have a spirit as indomitable as that of her master - who was now referred to in the press as Captain 'Stay-put'.

At last, some time towards the evening of that hellish day, the gale moderated and they were able to get under way once more. But the respite did not last for long. At 3 am, when Carlsen and Dancy were asleep in their respective corners, the sea worked itself up into another frenzy and the tow line snapped. Six blasts on the *Turmoil*'s whistle signalled the bad news. The two men hurried on deck. They were now fifty-five miles from Falmouth. Was this the ultimate disaster - the wretched end of all that stubborn toil ? Neither man would have admitted the fact ; but, at that moment, it seemed horribly probable.

Nothing could be done until dawn. Before another line could be sent across from the tug, the debris of the last effort had to be cleared away. The hardest part was that of removing the huge shackle. It took several hours of hard work with a hacksaw. When they were done, it was too dark even to think about sending a hawser across. The night closed in ; perhaps they would manage it next day.

Life seemed to be a succession of 'next days'. Would they ever get the *Flying Enterprise* to shelter ? The captain of the *Willard Keith* obviously had doubts about it. Shortly after dawn (on 10 January), he radioed 'A helicopter is standing by to take you off.' He clearly did not know the measure of Carlsen, who replied, 'I'm not abandoning.' There was no more to be said. The cargo of pig iron had acted as ballast and prevented the ship from capsizing. Nevertheless, she now appeared to be appreciably lower in the water, and she was lying over at a more perilous angle than ever. As the captain of another tug that had turned up reported, 'The *Enterprise* is flat on her side.' But Carlsen remained hopeful. After thirteen nights in his uncomfortable quarters, of facing day after day of continuing anxiety, and after suffering the worst excesses of the weather, he still refused to admit defeat. With Falmouth so relatively close, he believed he might accomplish the rest of the trip by allowing his ship to drift. Captain Parker's comment was, 'We can but hope and pray, hope and pray.'

She was drifting at $1\frac{1}{2}$ knots. At this rate, it would be the better part of two days before she reached the approaches to Falmouth. It was, in a way, a race against time ; for, to anybody who looked at her closely, it seemed unlikely she would be able to remain afloat for as long as that. Water was now pouring into the radio cabin ; Carlsen and Dancy had been driven out of their quarters and forced to find shelter in the chart room. As for the wind, it showed no signs of abating. At 2 o'clock on the afternoon of the 11th, the helicopter that had been hovering over the ship was compelled to return to its base. That night, Carlsen wrote in the log :

> She was a very well-built ship. I had command for over three years and crossed the Atlantic many times, so I knew what she could take. The last few days' gale was too much. There were high seas going and they were too much . . . Dancy and I agreed we would stick together and that when we had to leave, we would swim tgether, so it would be easier to pick us up.

The log of the *Flying Enterprise* had become Carlsen's personal journal. He confided things to it that he might not have told anybody. At last he was admitting to himself that the end of his ship was not very far away. No doubt he had discussed it with Dancy, for the two men had obviously achieved an ideal relationship. Even if the ship went down immediately, they would have accomplished what almost everybody else would have considered impossible.

Next morning, Parker reported, '*Flying Enterprise* listing badly. Port wing of bridge now under water. Standing by close to life-saving appliances in final degree of readiness.' And, later, to Carlsen, 'Your hatches are awash and may give way at any time.' The wheelhouse doors were now shattered, and water was coming in through the smokestack. Now, at last, they were compelled to surrender. Holding on to a lifeline, Carlsen and Dancy walked up the side of the funnel. They were obviously going to jump. Parker radioed a destroyer from the Royal Navy that had joined the group, 'I am going in for them. Stand by me in case of trouble.'

They went into the sea together, and together they swam the short distance to the tug. Dancy climbed up a rope ladder on the side of the *Turmoil* ; Carlsen was hoisted aboard. The whole business had taken less than ten minutes. Captain Parker transmitted, 'We have got both off *Enterprise*.' The *Willard Keith* replied with 'Congratulations !' ; but, by this time, Parker was entertaining the two men to tea in his cabin. Said Carlsen, 'I have done my best ; it could not be helped.'

By mid-afternoon, ninety per cent of the *Flying Enterprise* was under water ; only her starboard bow could be seen above the waves, and the sea was littered with her cargo. And yet, even now, this brave ship put up a struggle. She seemed to come up and then to go down again, to come up and to go down. But the life force, or whatever it is that exists in a ship, was spent. For an instant, she seemed to shiver. Then, silently, almost shyly, she slipped to the bottom of the Channel.

Carlsen, much as he may have disliked the idea, had become a popular hero. He received thirty three awards for his courage ; Hollywood made him an offer – newspapers competed against one another for his story, but he turned them all down. 'I don't want a seaman's honest attempt to save his ship used for any commercial purposes, or to get anything out of it,' he said. 'I only did what was my duty.' His reward came in the following year, when he was put in command of another ship. Her name was *Flying Enterprise II*.

11 HAS ANYBODY SEEN OUR BOMB?

Thirty years after the end of the Second World War, it was still possible to come across an unexploded bomb in Britain (and, no doubt, in Germany, too). Any such discovery was treated with suitable concern. Homes in the vicinity were evacuated ; the army experts moved on to the scene and defused the devilish device. Suppose, on the other hand, that it had gone off. There would have been considerable damage and a number of people would have been killed. But, nevertheless, the tragedy would have been local. The situation would have been beyond the dreams of even the most ambitious terrorist. On its own, however, it would not have been sufficient to wipe out a town - let alone a city.

But, ever since the original atomic bomb was produced in 1945, the technology of destructive hardware has taken on a new dimension. The war lords have learned to think big. There is no need to rest content with destroying part of a place ; the whole town, city, county or (depending on its size) country can be removed - with economy - and without exception. Indeed, with a little effort, it might be possible to strip the world itself of life (with the possible exception of rats, which seem to have an unusual talent for survival). Even in 1966 - a paltry twenty-one years since high explosive bombs had become, to all intents and purposes, obsolete - the armourers of the United States were producing 20-megaton nuclear devices. They were, the Pentagon assured anybody who was likely to be impressed, one hundred times more powerful than the bomb that had been dropped on Hiroshima. As an advertisement for American achievement, they were, perhaps, less appealing than Coca Cola. But, with the Russians manufacturing engines of extermination on a similar scale, what was one to do ?

The thought of these fiendish weapons kept carefully in store was disturbing enough. The very fact that they *existed* seemed to make life on this unquiet planet a little bit more hazardous. They were unlikely to explode because somebody had carelessly thrown away a lighted cigarette. But, with monsters such as these, the public felt uncertain. For sure, there were fail-safe gadgets, every precaution was taken, nothing could go wrong – there were plenty of reassurances. But there were some who believed that World War I had been started by accident (according to one theory, it might have been averted, if the German railways had been equipped with a more efficient signalling system). If *that* had been an oversight, might not World War III begin by error ?

What was even more disquieting to anybody who had a reasonable enjoyment of life was the thought that some of these nuclear bombs were not kept, so to speak, in the fridge. They were actually flown about the world in bombers. Again, there were reassurances. They were, the military experts hastened to explain, 'unarmed'. Just so ; but the more tiresomely sceptical members of the earth's community were apt to wonder what the point was. To employ powerful aeroplanes to cruise through the skies carrying weapons which, in the last resort, would fall to earth with a dull thump, must surely be to waste a great deal of money. They had, they felt, heard such things before. One day, they observed darkly, there would be a moment of truth. Would they survive it to say, 'I told you so' ?

Such a moment occurred on 17 January 1966. An aircraft carrying four H-bombs had taken off from a base in North Carolina. At about the same time, a KC135 aerial tanker departed from an airfield in Texas. The two were due to meet somewhere in the sky over Spain – where the bomber (which was, it was stressed, on a purely routine flight) was scheduled to replenish its supplies of fuel. Such encounters in the air were commonplace. The art had been perfected over many years. They were as unremarkable as the flight itself.

On this occasion, however, somebody got it wrong. Instead of gently mating ; instead of the KC135 refertilizing the bomber and then withdrawing, the two aircraft collided. Seven members of the two air crews were killed. The remains of the aeroplanes were scattered over a sizeable area of Spanish countryside. Among them were three of the four H-bombs. They were located without any great difficulty in a river valley, and discreetly removed. If it had not been for the fourth member of this unholy quartet, the matter would probably have ended there. It might even have been kept out of the newspapers. In a rather bleak sort of way, it was, perhaps, encouraging. The bombs had fallen to earth from a height of 30,000 feet, and not one of them had exploded. When the Pentagon had said they were 'unarmed', they had really meant it.

However, the problem of the missing bomb remained. As a character in *The Importance of Being Earnest* remarks, 'To lose one parent may be regarded as a misfortune ; to lose both looks like carelessness.' Something similar might be said to reflect the world's attitude to a missing hydrogen bomb. The Spanish authorities immediately protested that US aircraft carrying nuclear devices were not to fly over their territory. The Russians chipped in with the observation that, if the US Air Force persisted in making flights with atomic weapons on board its machines, the excursions should be restricted to American air space. The British Government was more restrained in its comments. It contented itself with muttering something about the danger of contaminating the high seas.

To their credit, the Americans acted very promptly. Within two days of the crash, a thousand troops belonging to the USAF were carrying out a meticulous search. They collected the wreckage, and located the three bombs that had fallen on dry land. They were found in the dry bed of the Almanzora river. The impact of coming to earth had split their cases, and fragments of plutonium and uranium had been scattered over the surrounding fields. Although the fall-out was described as 'moderate', it was sufficient to write off the local farmers' crop of tomatoes. They estimated the loss of revenue at £138,000. No doubt their claim was justified, for the soldiers eventually filled 4,810 airtight drums with soil, and dispatched them to the Savannah River burial ground for nuclear waste in South Carolina.

When the search was over, there was still one bomb that had not been accounted for. The four airmen who survived the collision had been picked out of the Mediterranean by fishing boats. Three bodies were also recovered from the sea. Here, then, must be the resting place of the truant bomb. The US Navy's Task Force 65 under the command of Rear-Admiral William S. Guest was assigned to find and retrieve it.

Was there any danger that, unlike the other bombs, it might explode ? The chances of such an occurrence were dismissed as 'infinitesimal', 'impossible', and 'one in a million'. Since there is a difference between 'one in a million' and 'impossible', the world's fears were not entirely removed. Nor after the experience in the Almanzora river bed, could the possibility of the sea's contamination be discounted. As if to cap this list of anxieties, there was also the hazard that it might fall into the wrong hands. The odds against it were enormous, but it had to be considered.

Rear-Admiral Guest had a difficult job on his hands. In a moment of candour, he told newsmen that he would have preferred combat 'any time'. Since the Admiral had been the first aviator in the Second World War to sink a Japanese ship, his remark carried conviction. On the other hand, the US Navy was obviously going to do all it could to help him. It would not be a simple matter of sending down frogmen ; something more technically sophisticated was required. As good fortune would have it, the navy had an assortment of submarine craft – some of which were very advanced indeed. The trouble was that at least two of them were in the United States. They were too big to be flown out to Spain, and the journey by ship would take at least a week. The news that they were on their way was released on 29 February.

By this time, sixteen ships of Task Force 65 were scouring the sea bed in an attempt to find the bomb. The first clue came from a fisherman living in the village of Palomares. He had, he said, seen what looked like half a human body attached to a parachute enter the sea at a point about six miles out on the 17th. When he was questioned he agreed that he might have been mistaken ; it could have been the missing bomb. This, at least, narrowed down the area to be searched – though the nature of the underwater landscape did little to help. It seemed to be an extension of the topography on shore – a submarine range of mountains. Had it been flat, there would have been few problems, but the folds and canyons provided far too many hiding places. To make matters worse, avalanches seemed to occur. An object that was visible one moment, might, within minutes, become buried beneath several feet of mud.

One of the men who went down in a midget submarine was diver Jon Lindbergh – son of the famous flier, Charles Lindbergh. After making sixty dives and spending 120 hours underwater, he was compelled to admit failure. The main difficulty, he said, was the litter of outcrops on the sea floor. All told, 260 items were recovered from the bed of the Mediterranean during the initial search period. None of them, unfortunately, was a hydrogen bomb.

The vessels in transit from the United States were named the *Alvin* and the *Aluminant.* The former was a seventeen-ton, twenty-two-foot-long craft produced by Litton Industries for the Navy's Oceanographic Institution at Cape Cod. It was manned by two men, though sometimes a third travelled on board as observer. In the whole of the USA, only three submariners had been trained to pilot her. On this occasion, she was to be handled by William Rainnie, a graduate of the Naval Academy. He was to be accompanied by a chief electrician and submarine electronics expert named McCannis. McCannis had previously served in the nuclear submarine, *George Washington.* The two men had already taken their unorthodox craft to a depth of 6,000 feet in an underwater canyon off the Bahamas.

In 1960, a bathyscope name the *Trieste* had descended to a depth of 7 miles in the Pacific. So far as the present search was concerned, the use of such a vessel was impracticable. It was too slow and, horizontally, its range was limited to short distances. Nor did it seem to be necessary. Three miles offshore from Palomares, the sea reached a depth of 600 feet. At this point, however, the underwater picture changed dramatically. It was as if the edge of a plateau had been reached. Suddenly, there was a sheer drop of 1,200 feet and then, mile by mile, the descent continued. Ultimately, the slope came to an end at 18,000 feet.

Admiral Guest had four deep-diving vessels in his task force. They included a 'deep jeep' – a two-man craft that had gone down to 2,000 feet without any misadventures, but which would be far beyond its range at 18,000 feet. The *Alvin* would have been literally out of its depth, but the *Aluminant* was an entirely different matter. On 11 November, 1965, she had set up a record by going down to $1\frac{1}{4}$ miles, and this was nothing like her full potential. She was the only submarine in the world designed to dive 3 miles. Painted in red, orange and blue, she was fifty-one feet long and weighed seventy tons. She made her journey to the Mediterranean in LSD (landing ship dock) *Plymouth Rock.*

By now Guest had the most sophisticated diving apparatus in the world at his command. Even more ingenious than the submarines, however, was an extraordinary piece of machinery known as CURV (for 'controlled unmanned recovery vehicle'). Directed from the surface by means of closed-circuit TV, it was surprisingly agile and could perform reasonably complex tasks. The idea was that the submarines, probably the *Aluminant,* would locate the bomb; CURV, if conditions were suitable, would bring it to the surface.

The *Alvin,* more commonly referred to as the 'Chipmunk', had reached the scene first and was promptly put into service. On each dive, she covered five miles of sea bed – travelling at speeds of between 4 and 6 knots. Rainnie received his instructions from surface craft by sonar telephone.

What strange journeys they were! In many ways, the landscape was like a reproduction of the land mass of Spain itself; granite mountains, in some places split by crevasses – in others penetrated by mysterious caverns. Now and again, a landslide set up turbulence and the *Alvin* shuddered. The small submarine was now five miles off the coast, cruising at 2,500 feet. Outside her hull, it was pitch dark; with her searchlight on, the range of vision was only thirty feet. But it was enough to reveal the nature of this eerie world beneath the sea – a realm that had never before been seen by a human eye. Was this what the earth would have been like, if a succession of hydrogen bombs had wiped it clean of life: nothing but a chaotic jumble of mammoth rocks swathed in absolute darkness? Was the missing H-bomb lying in the kind of setting it was designed to produce? Perhaps it was an interesting illustration of cause set against effect.

But the crew of the *Alvin* were practical men who wasted no time in such reflections. They had only one preoccupation - to miss nothing that was illuminated by the shaft of light ahead of them. For a moment, it seemed as if there were something that answered to the bomb's description. The submarine paused, hovering over the spot. Then it moved on. It was just another disappointment, useless object number 260.

Ahead of them, there was a scar across a slope which, Rainnie estimated, fell from the invisible heights above at an angle of seventy degrees. Now there was something else in front of them. It looked like the remains of a parachute - the sort of thing that is attached to a hydrogen bomb to slow down its descent to earth. Once again, the *Alvin* glided to a halt and, turning her two propellors outward, paused over the spot. It was definitely a parachute; nearby, there was a long black shadow. Was this it? The date was 17 March - precisely two months after the crash had taken place. The submarine crew photographed the object; there was a quick exchange of messages over the sonar telephone - Rainnie was instructed to surface. The parachute was definitely a clue. Its importance would become clear when the pictures were developed.

The previous month had produced little but worries for Admiral Guest. An electrical fault in the *Aluminant* had delayed her use. On 13 February a Globemaster 124 aircraft, carrying stores from the United States, had gone missing with a crew of eight somewhere between Seville and the Mediterranean Coast. It was thought to have crashed in the Sierra Nevada, but high winds had hampered the search. The bad weather persisted into the following month; on 2 March, it became so bad that operations had to be brought to a standstill. Operation 'Broken Sword', as it was called, had been going on for over six weeks, and it had accomplished very little.

In Madrid, the Spanish authorities were beginning to share Admiral Guest's concern – though for a different reason. The Spanish economy depended considerably on tourism. With an H-bomb lurking in the depths off the Mediterranean coast, the trade was bound to suffer. The sea might just as well have been populated by man-eating sharks. Something had to be done to convince the public of Europe that, unpleasant though the intruder might be, it was harmless.

Biddle Duke, the US ambassador to Spain, volunteered for the demonstration. Accompanied by his family and by the Spanish Minister of Information and Tourism, he travelled to Palomares – where, before an audience of press photographers, the party went for a bathe. 'It was not,' said Mr Duke, 'a jest or joke.' Considering that the day was cold and a strong wind was still blowing, nobody doubted him. The important thing was that the bathers came ashore afterwards with no ill effects.

On her return to the surface, the *Alvin* was hoisted back to her parent ship the LSD *Fort Snelling*. The navy oceanographic ship *Mizar* was ordered to a point on the sea above the submarine's discovery. Now, there was nothing to do but to wait until the photographs had been developed and the experts had examined the prints. Presently, the task force heaved a collective sigh of relief. The parachute's clue was not false; a few yards away, there was undoubtedly the twelve-foot-long shape of an H-bomb.

It was good news; but it was neither the end nor the beginning of the end. It was one thing to find a large object $1\frac{1}{2}$ miles beneath the surface of the sea; it was quite another to bring it up to the surface. Nobody had ever tried to accomplish such a feat before. It was more than difficult; it might be impossible. It would, perhaps, have been easier, if it had been a living enemy; for, where there is life, there is movement – and where there is movement, there are sometimes mistakes. An adversary leaves an opening, mistimes a movement, lowers his guard for an instant. Then, if one reacts quickly enough, there is an opportunity. The very mind that gives a man his strength also betrays him and creates his weakness.

But an H-bomb has no mind. It is obscene; it contains the kind of energy that belongs to the sun – and yet, until it is exploded, it is entirely passive. The men who employ it may make mistakes, but an H-bomb never does. Lying on the bed of the ocean, it is just a lump of deadweight; an inert assembly of components contained in a metal case.

Alvin and *Aluminant* went down together in an attempt to work out a salvage procedure. The *Alvin* was fitted with a mechanical arm – complete with 'wrist', 'elbow' and, at the point where it was connected to the hull, a 'shoulder'. It was intended to use this to attach a three-inch nylon line (with a breaking strength of 21,000 pounds) to the *Mizar*'s anchor chain. Once this had been accomplished, the other end of the line would be attached to the parachute. At first, everything went reasonably well. Then, just as the *Mizar* was beginning to haul up her chain, the line snapped. The bomb fell back on to the sea bed and slithered down towards the foot of the underwater mountain. On its way, it churned up the mud until, by the time it had reached the bottom, it was completely covered.

Another week went by. During it, the *Alvin* made thirty-four dives, spending a total of 250 hours under water. At last, on 2 April, Rainnie again saw the parachute. It was sticking out of the mud at a depth of 2,850 feet. The only good thing that could be said for it was that, in this new location, the slope was more gentle. Thirty-five degrees, Rainnie estimated.

For Rainnie and his colleague, this was the worst time of all. They tried to repeat the performance of the submarine's arm and the nylon rope; but, on this occasion, the parachute flared out. It threatened to wrap itself around the *Alvin* ; they managed to escape only by taking very prompt evasive action. To make matters worse, they were now informed that a storm was raging on the surface. It would be impossible for the *Fort Snelling* to pick them up. They had better stay where they were.

The two men remained in the belly of the sea for twenty hours before they were able to surface once more. During this long vigil, they made some interesting discoveries. At this depth, it seemed, everything is much bigger. Fish that looked as if they weighed between 500 and 800 pounds swam through their beam of light.

Even more improbable were shrimps that were at least one foot long.

For the next few days, the *Alvin* and the *Aluminant* took it in turns to watch over the bomb. Since there was a limit to the endurance of their batteries, they had to relieve each other at reasonably regular intervals – though the *Aluminant* once remained under for twenty-four hours. Eventually, it was decided that the *Alvin*'s 'arm' was not suitable for the work. That giant robot named CURV would have to be employed. It had already shown its value by recovering torpedoes off the coast of California. During these missions, however, it had never operated at a depth of more than 2,000 feet. Would the pressure farther down be too much for it?

CURV operated from a submarine rescue ship named the *Petrel*. The machine had been invented by an officer named Robert Pace of the US Naval Ordnance Testing Station at Pasadena. Its every move could be watched on a TV screen aboard the vessel; its actions were controlled from a console. Basically, it had an arm and a four-pronged grapnel with 3-inch-circumference nylon rope attached to it. Once it had placed a grappling hook in position, it came to the surface and was loaded with another. This process could continue until there were enough hooks in position to make lifting feasible.

At last matters seemed to be improving. CURV was lowered into position and sent towards the parachute. The hook was placed into position, and the robot was brought back to the surface. The second attempt was equally satisfactory. At 5.15 on the morning of 7 April, the machine went off on its third mission. If it got it right this time, the bomb could be hauled up.

But the supply of luck had run out. On this occasion, the parachute repeated the performance that had brought the *Alvin* so close to disaster. It was almost as if, as the more lively member of the bomb/parachute partnership, it was determined to keep its companion at the bed of the sea. Somehow, it managed to enshroud CURV – rather as a fisherman might net a fish. It took two hours to free the robot; but, in the end, they succeeded. The only thing was that, instead of three grapnels, there were only two in position. The combined breaking strain was estimated at 21,600 pounds. Was it sufficient, or should they make yet another attempt? Guest had to make the decision; he decided to chance it. Slowly, laboriously, the long haul to the surface began.

At two minutes past seven, it became clear that the weight was unevenly distributed. Nevertheless, the cables were holding. That was the main thing; and, as the lethal load neared the surface, frogmen went down and attached more lines. At 8.45 on Sunday 17 April, the truant bomb was hoisted on to the deck of the *Petrel*. The job of raising her had taken one-and-three-quarter hours.

On the following day, a party of newspaper reporters were taken alongside the *Petrel* in Admiral Guest's flagship, the US Navy's guided-missile cruiser *Albany*. It was the first time that a modern nuclear weapon had ever been shown to the press. One of them described it as 'shaped like the silver foil container for a good quality Havana cigar'. It was slightly dented and there were some barnacles on it. Otherwise, it seemed none the worse for its 30,000-foot fall, and the fact that it had been submerged for eighty days.

But it was, perhaps, an anticlimax. It seemed such an unimportant-looking object to have caused so much trouble; such an unprepossessing package for an instrument of extermination. No doubt the scientists responsible for it might have remarked that elegance was not a requirement for such a weapon. Nevertheless, it seemed inappropriate that the world might be ended by something that could only be likened to a discarded cigar container.

12 TRIBUTE TO OLD LIFEBOATS

When Grace Darling and her father rowed out to the *Forfarshire* in 1835, they acted, so to speak, as freelances. A few miles down the coast at North Shields, there was a lifeboat station, where Grace's brother was a member of the crew. Officially, *they* should have gone to the rescue of the steamship's survivors, but the weather seemed to make such a trip impossible.* The Darlings were in the habit of reacting promptly to distress signals. There was no sense in hoping that the men from North Shields might reach the wreck; consequently, when they saw the steamer's remains, they took matters into their own hands.

By a coincidence, the first lifeboat station in Britain had been established during the latter part of the eighteenth century at Grace Darling's birthplace - Bamburgh in Northumberland. It was the result of a legacy from the third Baron Crewe, Bishop of Durham, who died in 1721. Known as the Crewe Trust, the chief administrator was Dr John Sharp, Archdeacon of Northumberland. Dr Sharp's office was in Bamburgh Castle.

Presumably taking the view that charity should begin as close to home as possible, Dr Sharp set about organizing a means of rendering help to shipwrecked mariners. By 1771, Bamburgh had become well equipped to handle such responsibilities. The boats were ordinary craft - used for most of the time by local fishermen. In 1786, however, Dr Sharp was introduced to a carriage builder named Lionel Lukin. The meeting was of profound importance. For the first time ever, there was talk of constructing a boat

*As mentioned in Chapter 1, the lifeboat did manage to go to the wreck, but the Darlings had got there first.

especially for life-saving purposes.

Lionel Lukin had premises in Long Acre, London. For some time, he had been experimenting with the problem of constructing what he called an 'unimmergible' boat. The Prince of Wales (later George IV) had become interested in the venture and offered to pay Lukin's expenses. The Admiralty and Trinity House, both of which might have been regarded as likely customers, were not impressed.

When Dr Sharp inspected Lukin's handiwork, he decided it was on the right lines. The coach builder had purchased a Norway yawl, which he had proceeded to modify. He added buoyancy tanks to it; fitted a heavy iron keel to increase its stability; and attached a projecting cork gunwale to the top. At the waist of the boat, the gunwale was nine inches thick – tapering at the bows and stern to virtually nothing.

After their meeting, Dr Sharp sent Lukin a boat to be given the unimmergible treatment. Records are vague about its achievements, though it seems to have saved a number of lives during its first year of service.

Bamburgh now had a lifeboat station and, even more to the point, a specially built lifeboat. The rest of the country had nothing. The inadequacy of such a situation was rammed home to the citizens in the area of Newcastle-upon-Tyne on a stormy day in 1789. Some time during the afternoon, a ship named the *Adventure* was wrecked near the mouth of the river. She was only about three hundred yards off shore; the beach was thronged with hundreds of spectators, but nothing could be done to save the *Adventure*'s crew. Huge waves were battering the vessel and it was obvious that, before very long, she would break up. The seamen took to the rigging and waited without very much hope for somebody to come to their assistance. Nobody came. One by one, as cold and exhaustion took hold of them, they fell from their precarious perches and were drowned. There were no survivors.

The idea that men could die within yards of the beach – watched by innumerable bystanders – appalled the inhabitants of South Shields. The catastrophe had not been produced by indifference; the sorry truth was that, in the storm, it had been impossible to use any of the local boats. Perhaps they had heard of Lukin's craft which had done so well at Bamburgh; at any rate, they decided the time had come for South Shields to have a lifeboat of its own. A social club that styled itself the Gentlemen of Lawe House formed a committee and offered a two guinea prize for the best model 'calculated to brave the dangers of the sea, particularly of broken water'.

When the competition was announced, it attracted a great many entries. Eventually the adjudicators selected two as worth serious consideration. One, sent in by a gentleman named William Wouldhave, relied heavily on the use of cork. The other, submitted by Henry Greathead, had a curved (as opposed to a straight) keel. Since Mr Greathead was the only practical boatbuilder to take part, he was given the job of incorporating his own ideas with those of Mr Wouldhave and turning them into reality.

Henry Greathead's labours produced a boat that was thirty feet long. True to his own notion, the keel was curved; true to William Wouldhave's it was fitted with a cork lining, twelve inches thick, that extended from the deck to the thwarts. There was also a cork fender on the outside; all told, Mr Greathead used seven hundredweight of this material in the construction.

The result was extremely successful. Between 1791 and 1797, Greathead's lifeboat rescued the crews of five ships. Nevertheless, if he had expected a spate of orders, he was to be disappointed. Nobody outside South Shields showed any interest in his achievement until 1798. In that year, the Duke of Newcastle ordered one to be built at his own expense. In two years, it saved the lives of three crews, and the Duke's patronage seems to have succeeded where the worthy burghers of South Shields had failed. The contents of Greathead's order book grew more impressive - until, by 1803, he had built thirty-one lifeboats. Eight of them were for foreign countries. The Czar of Russia sent him a diamond ring as a present; Parliament voted him £1,200 and there were gifts of money from Trinity House, Lloyd's and the Society of Arts. But such rewards were well deserved. During 1802, one of Greathead's boats had saved the lives of two hundred people at the entrance to the Tyne, and there had been other, similar, rescues.

Despite Parliament's generosity to Henry Greathead, it seemed indifferent to the overall problem of saving lives at sea. If a community felt that it should have a lifeboat, its members could commission Greathead to build one. Nothing was done to create a system of lifesaving on a national basis until 1824. The instigator was a resident of the Isle of Man named Sir William Hillary. One of Sir William's friends was an MP named Thomas Wilson.

Hillary was that rare being, a visionary and a man of action. In the latter capacity, he had been largely responsible for rescuing sixty-two people when the steamer *City of Glasgow* ran aground in Douglas Bay. He had snatched eleven men from death when the brig *Leopard* foundered; nine, when the sloop *Fanny* became a total wreck. But his most heroic exploit took place when the mail steamer *St George* struck a rock and became a total wreck. During the rescue operations, he was washed overboard with three other men. When he was eventually hauled to safety, six of his ribs were found to have been fractured. Nevertheless, he had helped to save twenty-two sailors.

All this was much to Sir William Hillary's credit. But, as he was very well aware, there was a less satisfactory side to the picture. Over a period of twenty-five years, 144 vessels had been wrecked on the island's shores, 172 lives had been lost, and something like a quarter of a million pounds' worth of property had been destroyed. Statistics such as these caused Sir William to compose a pamphlet entitled *An Appeal to the British Nation on the Humanity and Policy of forming a National Institution for the Preservation of Lives and Property from Shipwreck.* It was published in 1823. His friend, Thomas Wilson, took up the cause; William Wilberforce became interested; and so did King George IV. In the following year, the Royal National Institution for the Preservation of Life from Shipwreck (usually abbreviated to the 'Shipwreck Institution') was formed. Some years later, its name was changed to The Royal National Lifeboat Institution.

The Shipwreck Institution began its life with £9,800 from which, among other things, it paid for a dozen boats. It would be nice to record that, thereafter, it became more and more affluent. Such, however, was not the case. During the following year, its receipts slumped to £3,392.37½ ; and, eleven years after that, they amounted to a paltry £250. Did this mean that it had failed ? On the contrary ; if one remembers that it was formed to save lives, it had already shown itself to be triumphantly successful. The parsimonious Victorian public may have been loath to part with its pennies. Happily, the fishermen around the coast, on whom the Institution depended for most of its crews, were less reluctant to offer their lives. Now and again, after a particularly spectacular rescue, one lifeboat crew or another received extra rewards, but this was a matter of chance. Time and again, they went out in the very worst of weather, putting their lives at risk to save, sometimes, a mere handful of sailors from the perils of the sea. In the middle of the nineteenth century, their only remuneration was 35p for a daytime rescue ; 50p at night.

If the Institution's second annual report fell short of an accountant's dream, the falling funds were more than outweighed by the record of 342 lives saved. This, surely, was the real measure of its early achievement. But sometimes it experienced tragedies of its own. In October 1841, one of the boats at Blyth in Northumberland capsized and ten men were drowned. In February 1843, the boat at Robin Hood's Bay, Yorkshire, was taking off the crew of a London vessel named the *Ann*. Several men jumped into her at the same time as a rogue wave struck her. She capsized ; three were trapped beneath her – the others managed to clamber on to her bottom. Presently, in her upturned state, she drifted towards the shore. But the violence of the storm was not yet done. As she approached the beach, a giant comber turned her end over end. All told, one officer of the coastguard service and eleven men were killed.

On a day in December 1849, the South Shields lifeboat went to the aid of the *Betsy* – a vessel from Littlehampton that had gone aground on a sandbank. They were about to take the survivors on board when two waves, working with deadly co-ordination, threw the boat over. Twenty-three out of the twenty-four aboard her were drowned.

But such occurrences, fortunately, were the exception. In the year following the *Betsy* disaster, the service was reorganized. Qualified inspectors were appointed to ensure that the boats were maintained in a seaworthy condition, and an improved scale of payment was established for coxswains and their crews. Considering the hazards of their calling, it was ridiculously small. The former received £8 a year - the latter, £2. When the boat was called out in the daytime, they each received a bonus of 50p ; if it happened at night, the sum was £1.

Over half a century had gone by since Henry Greathead had designed the definitive lifeboat. During this period, a number of new ideas had been conceived ; some of them, possibly, might benefit the rescue service. Once again, the reigning Duke of Northumberland showed himself to be not only enlightened but also munificent. His Grace offered a prize of 100 guineas for the best model of a new type of boat, and another 100 guineas towards the construction of the winning entry.

Two hundred and eighty ideas were submitted. They came from all parts of the UK, from France, Holland, Germany and the United States. After six months' cogitation, the adjudication committee decided to give the award to a Mr James Beeching of Great Yarmouth. Mr Beeching's achievement was to establish a principle that resulted in the self-righting boat. It involved the use of a heavy iron keel and air chambers ; features which, when the craft was upside down, made her unstable. From time to time, the designers of big ships have misplaced the centre of gravity, producing vessels that, in a storm, were liable to capsize. Considered from the viewpoint of a man standing on his head, however, these awesome miscalculations become virtues. If a boat with her bottom upward is uncomfortable in this posture, she will do her best to turn over yet again - thus placing herself right side up.

Mr Beeching's idea seemed to be a sound one. By 1872, fifty-six per cent of the Institution's fleet was composed of self-righting boats.*

*Mr Beeching may have gone some way towards producing the ideal lifeboat, but he had not gone far enough. The very things that made them self-righting also, in the opinion of many crews, made the boats more liable to capsize. Consequently, they were unpopular.

The shores of the UK could now be said to be adequately covered by the lifeboat service. Some stations, obviously, were more severely exercised than others. The crew of the Ramsgate boat was particularly hard pressed, for the town looked out across the northern boundary of the Goodwin Sands. Elsewhere, wrecks were unusual events ; on the Goodwins, they were commonplace.

Ramsgate's lifeboat did not function on her own. Her partner in these errands of mercy was a tug named the *Aid* owned by the harbour authority. The idea was that the *Aid* would tow her to within the vicinity of the disaster ; after that, she would proceed by sail or by oars. During a spell of bad weather, the tug used to maintain a constant head of steam in readiness for an emergency.

On a February night in 1872, the Spanish brig *Samaritano* was on passage from Antwerp to Santander. On board were her master and a crew of eleven. The wind got up ; violent gusts churned up the sea, and sudden squalls of snow blocked out the visibility. The *Samaritano*'s captain misjudged his position, and the small ship was blown on to the sands opposite Margate. She was spotted by the crew of a lugger from that town named the *Eclipse*.

The *Eclipse*'s skipper envisaged the pleasing prospect of salvage money. If he put a crew on board the brig, it ought not to be beyond the bounds of possibility to refloat her on the next tide. He borrowed two men off a Whitstable fishing smack that was in the neighbourhood. Presently, eight seamen from Margate and the two Whitstable men were put aboard the *Samaritano*.

Whilst the times of high and low tide might be predictable, the weather was not. The rising water brought with it gale-force winds and high seas. The brig was now in serious trouble. One wave picked her up – and withdrew. Another repeated the performance, each time raising her up and then slamming her down on the sandbank. Before long, a hole had been knocked in her side, and she had begun to fill with water. The ship's boat was swept away ; the hatch forced open, and the cargo distributed over the unquiet surface of the sea. As she settled on one side, there seemed to be a possibility that she might turn over. To reduce the likelihood, they decided to cut down the main mast.

Before very long, the visitors from Margate and Whitstable had changed their ideas. There was no question of refloating the brig ; they were now fighting for their lives. The only safe place seemed to be the fore-mast rigging. In company with the Spanish crew, they swarmed up it. The wind and the waves did their best to tear them away ; and, as if to make the situation even more uncomfortable, the droplets of spray were freezing.

The first attempts at rescue came from Margate. The town had two lifeboats. In their hurry to get the first one away, the crew forgot to close the flotation chambers' valves. Although this caused her to fill with water, her crew brought her to within a quarter of a mile of the brig before they were compelled to turn back. The second Margate boat fared no better. A wave crashed on to the tiller and snapped it off. Eventually, she was driven ashore about a mile up the coast.

A pair of luggers now made an attempt at rescue. A squall removed the fore-mast of one within a hundred yards of Margate pier. The other got to within sight of the wreck, but was unable to get close enough. She, too, had to turn back. Everything now depended on the Ramsgate lifeboat and its indomitable companion, the *Aid.* The tug's master, Daniel Reading, and the lifeboat's coxswain, James Hogben, had worked together on several occasions. It was just as well ; within the next few hours, both men were to experience some of the most daunting moments in their respective lives.

A crowd had assembled on the pier. As the *Aid* went by with her companion in tow, the specators watched waves breaking over the small steamer. As one of them put it, 'She seemed to go under water. I would not have gone out in her for the universe.' The wreck was between ten and twelve miles away. A 300-foot length of hawser as thick as a man's wrist united the lifeboat with the tug. It seemed to be strong enough in all conscience ; but, as they neared the sands, it broke. Hogben had a notion that they could reach the brig by rowing. He was sadly wrong ; after a few minutes of back-breaking work, the men found that they had not advanced one foot nearer their objective. The best that could be said was that the exercise had warmed them up.

Somehow, and goodness knows how they managed it, a new line was passed across and the tow continued. The bursts of snow and sleet had begun like isolated raids ; now the tempo quickened and they seemed to be continual. The men in the lifeboat became so cold that they felt, as one of them put it, 'like a body of ice'. To make matters worse, and although the time was only a few minutes after noon, the visibility had become so bad that they could not find the wreck.

At last there was a break in the clouds and they were able to catch a glimpse of the *Samaritano*'s remains on the far side of the sands. Shortly afterwards, the hawser linking the lifeboat with the tug parted again. The rest of the journey would have to be made under sail.

By the time the lifeboat reached the *Samaritano,* the brig had been aground for eight hours ; Mr Hogben and his men had been at sea for four. One by one the more intrepid souls jumped from the rigging ; three of the Spaniards - one of whom was the cabin boy - had to be assisted. There were now thirty-two men in the boat. At some point on their way back to the *Aid,* the mast and sail were blown overboard ; but, beside all the other difficulties that had occurred, the job of recovering them seemed easy. When, at last, they regained the safety of Ramsgate, it was dark. But there were still people watching from the pier ; and, as the *Aid* and the lifeboat returned, they cheered. Later, the Board of Trade rewarded each member of the respective crew with a medal and £2. The Spanish authorities were more generous. They gave them each £3 and a medal.

When the emigrant ship *Fusilier* went aground on the Goodwins, the Ramsgate lifeboat crew rescued all her crew and passengers (including sixty women and children) by ferrying them, thirty at a time, across to the *Aid.* The ship was eventually refloated. During the following year, a Portuguese brig ran on to the sands. Her captain hoped that she, too, might be saved. It was, perhaps, understandable, but his determination very nearly caused the death of his own men and the loss of the lifeboat. It was the old story ; instead of improving, the weather worsened. When, at last, the men were persuaded to leave, they tried to bring their sea chests with them. The cumbersome boxes were swept away ; the brig broke up ; the lifeboat went aground ; the tug lost touch with her and patrolled the area of the sandbanks until dawn. When at last Captain Reading of the *Aid* gave up hope and put back into Ramsgate harbour, he found the missing lifeboat riding contentedly at her moorings. Her crew had not only managed to refloat her ; running before the wind with their small sail, they had got back ahead of the steamer.

The loss of their sea chests meant that the Portuguese crew were now bereft of all their belongings. Between them, the most they could muster up was £17. As the harbour mouth loomed into sight, a spokesman approached Mr Hogben. His companions, he said, were very grateful ; they would like to offer this money as a reward for saving their lives. Would the lifeboat crew accept it ? Mr Hogben was charming but firm. He and his men appreciated the offer, he said, but they would not dream of taking it. Heaven knows, they were not wealthy men, but he reckoned the stranded mariners needed it more than they did.

One of the crew members on this operation was a man of sixty. Afterwards, he said, 'Well, perhaps my friends were right when they said I hadn't ought to have gone out – that I was too old for this sort of work . But, you see, when there is life to be saved, it makes one feel young again ; and I've always felt I had a call to save life when I could, and I wasn't going to hang back then. And I stood it better than some of them, after all.' Nevertheless, he lost his voice afterwards, and was confined to bed for a week. It was hardly surprising. 'I cannot describe [the experience]', he said, 'and you cannot, neither can anyone else ; but when you say you've beat and thumped over those sands, almost yard by yard, in a fearful storm on a winter's night, and live to tell the tale, why it seems to me about the next best thing to saying that you've been dead, and brought to life again.'

The Ramsgate lifeboat was, perhaps, the last resort. One New Year's Eve during the latter part of the last century, the boatmen at Deal saw a ship go aground. Normally, she would have been seen by a nearby lightvessel, which would have signalled news of the wreck to the coastguard. On this occasion, however, a wall of fog stood in the way. The Deal men tried to put out in their luggers, but the weather was too bad. Making a collection of ha'pennies, they amassed enough money to send a telegram to Ramsgate. The post office made a mistake ; the message was delayed ; and, by the time the tug and the lifeboat reached the stranded ship, she had capsized. Twenty-eight lives were lost.

When the full-rigged ship *Providentia,* on passage from Finland, struck the Goodwins, the majority of the crew got away in one of the ship's boats and were carried across the Channel to Dunkirk. By the time the Ramsgate lifeboat arrived, only the master, the first and second mates, and an eleven year-old cabin boy were on board. The lifeboat was secured to the wreck by a line ; the captain tried to come down it, but a wave caught hold of him and threw him into the boiling sea. He was swept past the boat, but he managed to grab a lifebuoy thrown by one of the crew. Two hours later, he was picked out of the water – exhausted but still alive. The mate was less fortunate. He, too, was wrenched from the line but he was pulled under and drowned. Now there were two. One of the lifeboat crew managed to haul the second mate on board, but it was a near thing. He had to lean over the bows ; even with four men holding on to his legs, he was very nearly washed overboard. The cabin boy was fished out of the sea with a boathook. At first he was given up for dead. During the voyage back to Ramsgate, however, he recovered.

The Ramsgate lifeboat's score of successes during the last century was a substantial one. When, in 1881, the barque *Indian Chief* went aground on the Goodwins in an easterly gale, twelve of the twenty-nine on board were saved. The coxswain, Mr C. Fish, was awarded a gold medal – the remainder of the crew and the seven men who manned the tug *Vulcan* (which had now replaced the *Aid*) received silver medals. But perhaps the most impressive feat of a lifeboat crew during the Victorian period belonged to the men of Lynmouth in Devonshire. It took place during January 1899.

A ship was in distress in neighbouring Porlock Bay. A westerly gale was blowing at the time ; high seas were sweeping across the front at Lynmouth and into the harbour. Under these circumstances, it was impossible for the lifeboat to put to sea. What, then, should they do ? Give up all ideas of a rescue and let the sinking vessel die without making an attempt to take off her crew ? Somebody had a better idea. If the boat was transported overland to Porlock, it should be possible to launch her.

It was, perhaps, a crazy suggestion, but there seemed to be no alternative. The distance was at least eleven miles, and all of it lay over rough country. The boat had to be transported from sea level to the clifftop (a height of 1,400 feet), and then down again. In places, the gradient was one in four-and-a-half. Banks had to be dug away to make the route wide enough ; trees had to be sawn down ; and gateposts removed. The whole thing was accomplished at night – with oil lamps, which were continually being blown out by the wind, as the only illumination. Aided by a band of voluntary helpers, it took the crew ten and a quarter hours before they reached the beach at Porlock. They were dead tired ; and, throughout the ordeal, they had not paused to eat. Nor, on their arrival, was there time for a meal. They had to put to sea at once ; but they reached their objective just in time. The crew of the helpless vessel were removed to safety minutes before she foundered.

This, no doubt, amounted to some sort of record of endurance. The ultimate in this respect, however, probably belongs to the Ballycotton lifeboat crew in County Cork. During February 1936, the Daunt Rock lightship was in trouble. The boat put to sea in the teeth of a south-easterly gale, and spent the next sixty-three hours in the vicinity of the damaged lightvessel. On five occasions, she tried to go alongside and take off the crew ; on each, the fury of the sea forced her to retreat. More than once, it looked as if the storm's victim would capsize on top of the lifeboat, drowning everybody on board. But, at last, on the sixth attempt, they were successful.

During the years covered by this book, there have been considerable technical advances. Radar has been introduced, electronic navigation systems, automatic pilots, echo sounders, and so on. Two elements, however, have remained the same : the wind and the sea. If anybody succeeds in taming them, lifeboats may become unnecessary. Until then, there will continue to be a need for the Royal National Lifeboat Institution's 200 stations around the coast of Britain, and for the nearly £6,000,000 that it costs annually to maintain them. Rescues at sea are a matter of saving lives. That is what the RNLI is all about.